PATH TO ETERNITY

YELENA LOGINOVA

PATH TO ETERNITY

Practical Guidance and Case Study

Applications of Eternal Creative Harmonious Development

Human!
You are the World, you are Eternity.
You possess immeasurable powers.
Your possibilities are limitless.
You are the embodiment of the creator.
In you, His will resides,
Through his destiny you change the world.
In you, His love resides.
Love all life as He does, He who has created you.
Do not embitter your heart. Think good, do good.
Good will return to you with longevity.
Love will give immortality,
Faith and hope, prudence.
With faith and love
Your invisible powers will come alive.
And you will achieve all that you dream of.
Immortality, it is the face of life.
Just as life is the trace of Eternity.
(Grigori Grabovoi)

I consider this book a textbook for all those who choose to study the teachings and practice of Grigori Grabovoi's Technology of Salvation. When this knowledge is implemented, it can help solve and eliminate problems in our lives! Therefore, you and your family live in optimal health and vitality, and most importantly, it provides macro salvation for all....in other words.... physical, mental, emotional and spiritual wellness for all!

Grigori Grabovoi created at least 300 lectures and many books that are gradually being translated into other languages. Due to the large volume of information that Grabovoi has shared over the years, which would make for an overwhelming text, in this book I have highlighted the most important segments of information and teachings from his clairvoyant principles presented. I am hoping that by doing this it will inspire the reader to want to seriously study this grand divine Source of wisdom. My purpose is to arouse your interest in the knowledge of consciousness, for you to understand on a deep level that we are all creators of our reality, and to use our amazing Spirit, Soul, and Consciousness that resides within all of us to reclaim universal wellbeing for all of creation. If this world vanishes, then there will be no future for all. I also want to assure that the purpose of man is to live forever, think positively and take themselves as gods who are invincible with unlimited possibilities for creation. Most of what we have learned as children is a misrepresentation of the divine truth, leading us away from the harmonious world that was meant for us. You just need to decide to pull out from under the influence of the collective consciousness which holds a multitude of dogmas and distorted information which have been instilled in many since a very early age. Many of the things we have learned allows people to be manipulated by certain negative forces due to people not realizing the greatness of all mankind (including themselves), and therefore, keeping us in the dark about our goals and opportunities.

Grigori Grabovoi was born in 1963 in a village called Bogara in Kazakhstan, Asia.
From early childhood, he was remarkable for his ability to predict the future, as well as, to see the true tranquility and infinite nature of the human being and world in any situation. He has an amazing ability to think globally. He graduated in 1986 from Tashkent State University with a degree in mathematics and mechanics. He is an academician of the International Academy of Informatization, a Professor of Math, a member of the Russian Academy of Sciences, an Advisor of the Russian Federal Aviation Service, as well as, many other things. His name is included in the list of names that make up the intellectual elite of Russia.

Using the Technology of Consciousness, it has been proven that even diseases such as cancer and AIDS in the last stage can be cured. He teaches all people in the world to use of the Technologies of Salvation, and our goal is to learn to save ourselves and others. All lectures are presented on the spot. In other words, Grigori Grabovoi never prepares his new lectures in advance which can be complex in nature due to his great intelligence coming directly from the divine source. This is not simple information for even a well prepared person to understand, but his devoted students are working on clarifying the teachings so they are more comprehensible to more people, and I am happy to say that the translations and clarification of material is coming along nicely.

Out of all of his lectures you will not hear any words like sin, slave, devil, "fear of God", "you are the sheep", etc. There's only a positive and profound respect for the freedom of man as a collector of experiences which can be transformed into wisdom. Any experience is valuable, but the time has come when the participation of our lives must move toward an Eternal and Harmonious progression.

All of his phenomenal abilities (including clairvoyance, teleportation, telekinesis, and many other) have been confirmed with seals from hundreds of different institutions and they have even been certified by notaries and representatives of the UNESCO and UN.
Grigori Grabovoi proved that by using the Technology of Consciousness, even diseases such as cancer and AIDS in the last stage can be returned to the norm of the creator. He teaches his Salvation Technologies to anyone willing to learn. It is our responsibility to chose to always do "good for all" with all our actions, thoughts and words, thereby, saving ourselves and others and living in peace in harmony.

Grigori Grabovoi has given a lot of unique and new information that humanity has not yet dreamed of. He teaches a tool to act as effectively as the Creator. In the lectures knowledge comes to him directly from the Creator. I was fortunate enough to attend at one of the lectures, and I had the impression that he was not talking like a man of today, but instead like a human from the future. Grigori Grabovoi's lectures are not simple, and even for a trained person to understand them the first time around is not easy. Thankfully, like I stated before, the followers of his Teaching are working to present the information in a more simple form for even beginners to understand.
This textbook on the Technologies of Salvation is for those who are open to a new science in which the consciousness of every man with his spirit and soul are the main tools necessary for building a most harmonious reality.

Yelena Loginova
President of the Research Center of ENLIGHTENMENT. USA

ABOUT THE TEACHING OF GRIGORI GRABOVOI

The Goal of the Teaching of Grigori Grabovoi is the Harmonious Salvation (100% health for all of creation) for all and the maintenance of an Eternal, Creative, and Harmonious Development (evolution or expansion). Grabovoi states that all of the methods of salvation, healing, and controlling situations are based on giving people "a unique knowledge obtained from God". This unique set of skills and knowledge are made clear in terms and concepts of modern science and theory.

The main goal of human life is to raise the level of consciousness.

In the education process you will learn management techniques through your own consciousness to:
- Restore the physical body and to harmonize relationships;
- Control and optimize future events;
- Perform control for the prevention of disease, accidents, disasters, ecological problems, weather and more;
- Apply the technology to restore plants, animals, water;
- Extend the management methods in the areas of health, education, religion, sociology, economics, etc.;

In the learning process you build awareness of vitality, invincibility and immortality. Perhaps it may be the first time you learn to make choices in constructing your own reality, as the Creator does.

Our most important task is to actually prevent a potential global catastrophe. Students perform the action to save everyone and prevent a global catastrophe, and through this action he progresses in solving his own personal problems. Human beings are Godlike and when you are living by the divine plan you come to feel joy and hope, as well as, a responsibility that comes along with it. Every person has Free Will. It is our privilege; a Divine Gift from God, but also a huge responsibility as well. We must know: NOBODY HAS THE RIGHT DICTATE to others. You can only transfer knowledge....and the person will make his own choice.

The New Science of Grigori Grabovoi is a new direction of knowledge, claiming the path of development only in the direction of the creative pristine development of the entire world. For the first time discovered and confirmed in practice, the fundamental laws of the creative control of any object information in the direction of salvation from destruction. For the first time it has been mathematically proven in practice that the change in perception can convert the information in all systems of reality. This is unique knowledge and is interpreted in terms of the concepts of modern science.

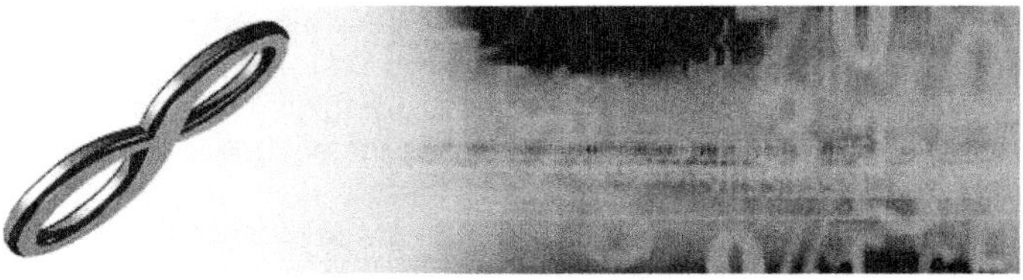

PROGRESS OF CONSCIOUSNESS

There is no need to torture your body with diets and medications. Instead, you must work on your mind. We are forming our consciousness with actions, thoughts, words, emotions, desires and passions. In general, Nature (and therefore, Man) is the information field. It is a collective consciousness that surrounds us and is within us. Everything in it is moving and vibrating. For instance, with influenza, information about the flu is in the collective consciousness. Since fear of the disease is found in the collective information field, it can "infect" a lot of people and cause many people to become sick if they do not chose to look beyond the collective consciousness.

It is necessary to form a new collective consciousness that can cure everything possible, including: cancer, AIDS, cirrhosis of the liver, missing or diseased organs and teeth, thereby, regenerating and optimizing the organs and cells of our bodies. The collective consciousness says that without food and water we can not live. If 70% of people claimed that we can live without food and water, then we would take the necessary substances for well-being from the air, water and sun.

It is important that the sick person himself send into the information field the belief that he has defeated his illness, internally and externally. It is also important for the people around him believed in it too. If this is done, every disease will be defeated with this process. To optimize the results, you need to have greater frequency and speed of thought. It is necessary to understand the procedure at the level of the fundamental laws of the universe.

Every person has the ability to heal and be healed instantly. The intention affects literally every cell in the human body. Intention creates emotion - a wave of energy, momentum or force that causes the intention in action.

"E-motion = Energy in motion = intention in action"

If you wish to become a true manager of healing, you need to have a healthy spirit, which you work with everyday. Each day, all of your 63 trillion cells are filled with happiness, beauty and health. Try not to be angry or get irritated. It is of utmost importance to forgive everyone

and clean up the arrogance. Constantly monitor your speech, thoughts, emotions, and passions towards doing good for all and to love everyone. Methods and technologies of Grigori Grabovoi will instantly be working under these conditions and there will be instant healing.
A sick person must send a strong impulse from the heart and soul for health and immediately all disease from head to toe will be removed. This is possible only if the healer has a healthy spirit.

The universe has a usual way of transferring knowledge and information between us, but when someone is not interested in spiritual knowledge, these technologies are hidden from us and interfere with our divine nature. Of course, this is an internal agreement within the person. Violence in our education system is completely ruled out.

From all this we can conclude:

* A person is not a small and closed part of the Universe, not a local structure. Consciousness is able to bring people to the limitless expanse of the cosmic manifestation, letting him feel the whole universe.

* We are completely determined by our consciousness: whatever is our consciousness so is the world around us.

* Scientists are right. By using our consciousness, we are able to create everything in the universe, including the immortality and an eternal, happy life in a space of love.

• With our imagination and thoughts, we can influence the body to improve health, as well as, manage and monitor our health and fitness. Everything in our life is controlled by consciousness and intention. The intention affects literally every cell in the human body.

• We constantly have to be managers and harmonize the environmental reality that is constantly controlled in the direction of the norm and harmony.

SOME QUOTES FROM BOOKS AND LECTURES OF GRIGORI GRABOVOI

Everything in the Universe is the *object of information*, and the whole world is a very complex information system of all living things on Earth are in the Information field as information objects. Life on Earth would be impossible if the living things would not have caught the incoming information from the environment, do not know how to process it, and use it to send information to other objects.

All objects in the world are related to each other informational connections. By changing any connection with, essentially, you can change the system as a whole. Academic Grabovoi created a mechanism control system, the essence of which is the impact on the information system of the human mind and through perception. Man is -an information object, the disease - also an information object, any situation around - a set of information objects and connections.

All information on the Universe is available in each point of space, contains in each object. The information can instantly change, it doesn't depend from space-time.

The Law of development forms of information. The world is constantly changing, therefore, any information developed.

If a person consciously combines his personal task to the global task of saving all, in accordance with the technology of his thinking becoming management.

Everyone on the personal experience of long-tested truth: how we think that we live in such a way. This means that a man of his thoughts, which are the vehicles of consciousness, builds the world, is building a reality.

Science rightly asserts the presence of consciousness ... in every electron, atom, molecule, cell of any substance.
The main task in life of human is to raise the level of his consciousness.

The basis of the construction of the World Creator put emotions and most of all joy, light and love. It is joy, light and love are the first foundation of the world. And they, along with proper understanding of the World will provide a full happy life.

Old ... slow mechanism of comprehension of higher truths and spiritual development through the rejection of the physical body, a temporary stay in another form on the subtle planes of existence, there gathering the right information and then return back into the physical body - this leisurely way to grow is no longer corresponds to the modern pace of life . So we should not give up the physical body, do not spend time on all these transformations, and in this very body with the help of special techniques to learn how to go into higher states of consciousness, the most secure spiritual growth.

In a canonical form the person has no illness. Illness is external information. Somewhere you have passed it in object of the information which corresponds healthy to you».

«Never fix negative information. Level prosperous dreams - avoiding the negative and catastrophic events. (Or «Never fix the negative information. The dream of good end of event helps to avoid accident or troubles.")

Knowledge of the laws and forms of governance - this is already to manage the physical reality. "

Soul - this is the substance that is created by the Creator in accordance with the eternity of the world and is an element of the World. The soul is unshakeable, in principle, it exists as an organizing structure of the world and therefore it is a reproduction of concepts such as, for example, the Spirit. Soul is part of the World, that is present in any event.
When you act on the tasks of your soul can happen an instant cure.

Spirit is the action of the Soul.

Consciousness - the structure that allows the soul to manage the body. The soul, the material part of which is the physical body, operates a

reality through consciousness.
In a broad sense, consciousness - a structure that unites spiritual and physical matter. Change of consciousness can transform the spirit, and thus reproduce action, then there is an event.
Personal consciousness of every human being is an integral part of the collective consciousness, which is of great energy power capable of managing these people.
The consciousness has light-golden color.

The person by his consciousness can influence time. The individual can enter its consciousness into the past or future and change there events to norm.

In the consciousness of each individual there is an area of absolutely pure consciousness, or it is possible to name like Divine consciousness. And through this divinity shown in consciousness, the person initially has all abilities and possibilities of God-Creator. The person has all knowledge of the God.
The physical body is the continuation of the Soul.

Thought - this is the information that is the link between consciousness, body and spirit.

The feeling - it is a manifestation of the soul in the dynamics of the spirit, and in the part where the spirit associated with the consciousness and body. When the soul is organizing body, the feeling is that the framework on which to build the body. The feeling when interacting with reality perceives the information faster than thought.

Time can be viewed as a transformer in a certain kind of space. In this case, the time can be thought of as a kind of force lines of the space over which the transformation and movement. If the time to understand the structure, it is possible at the right time in the right place to achieve the desired space, including the implementation of a desired event. The reverse option: changes in space, you can change the time. Both structures ... - consciousness and time are constructs of consciousness.

The space and time, is the construction of consciousness.

Clairvoyance is a universal method of access to information.

Enlightenment. An enlightened person - is the one who has seen the light of the soul. The soul of such a person has the light of knowledge, the light of the future, the light of creation. It is someone who has knowledge of creation, knowledge development, knowledge of harmony.

Close fraction of consciousness is our exact knowledge (rather than abstract), is information that we already have about any subject, action, situation, and so on. This is what we see, hear, feel around us, it just kind of information that we received from other sources (someone told, we have heard on TV and so on).

The distant fraction of consciousness is information which only can be received us, it is the information which we can have, but we have no, these are abstract knowledge of something.
For example, knowledge of any star from constellation the Alpha Centaurs is the distant fraction of our consciousness. Basically, it is that information which our consciousness still "has not highlighted".
Super far areas of consciousness provide super fast processing of the information. Therefore than you use farther areas of consciousness, you can carry out especially fast processing of the information».

We constantly have to be managers and harmonize the environmental reality that is constantly controlled in the direction of the norm and harmony.

Mathematics - the science which operate with endless rows, so the mathematics closer to reality.
The challenge is to transfer knowledge of the soul into a logical phase and use them consciously.

Of course, this is an internal agreement on the person.
Violence in our education system is completely ruled out.
Through spiritual growth reduced internal distortion

Collective consciousness is the unified consciousness of all people. There exist stable views of everything. Every man has his own ideas about everything differing from the point of view of other people. The resulting averaged view or conception- is a collective representation. And that it is perceived by people as an objective reality.

We must acquire a higher vibration, to go "easily" through the expansion of consciousness.
Within the collective consciousness of a man is just an element. Through the expansion of his consciousness, he comes out of the collective consciousness.
It needs to reprogram the mind on "the structure of the eternal. "

The man is always searching, but once he realizes he gets to the point of Eternity.

- The person has the false reasons of suffering. It is ridiculous idea about as if many things can be received only through sufferings. And now, in connection with threat of nuclear destruction, this idea about necessity of suffering can even lead to global accident ».
Actually for sufferings and negative emotions aren't present true base.

- It is the same with the other causes of false distress. Along with them will go a ridiculous idea, as if many things can only get through suffering. And now, because of the threat of nuclear annihilation, the idea of the need of suffering may even lead to a global catastrophe.

- The old mechanism of slow comprehension of higher truths and spiritual development through non-physical body (when dying), a temporary stay in another form in other planes of existence, get out there some information and then returning again to the physical body - this way to slow the growth of consciousness is already does not match the pace of modern life

- So we should not abandon the physical body, not wasting time on all these transformations, and in this very body with the help of special methods to learn how to go into higher states of consciousness, accelerating his spiritual growth.

- Management of a reality is carried out by consciousness through creation and optical perception of the images symbolically designating event or a situation or object of a reality. And by direct action in transformation of this optical image to norm, the physical object or event changes.

- Harmonization of any situation or health occurs through influence on these symbols structures of absolute norm – by Streams of Eternity, Love, and Light of the God. And also through the transformations made by the consciousness with the forms of government, or change of inharmonious combinations of color in harmonious.

- Gradually, the knowledge is activated in cells of the body of Eternity information, information self-regeneration, and rejuvenation. Human cells are made eternal. According to the laws of universal connection, even one such cell can restore the entire body. And when there are many such cells, the body becomes eternal.

- Eternal life - is a constant self-recreation.
- Offered technologies are carried out without pressure, in a quiet

condition, meditations here aren't necessary.

- If you have not yet clairvoyance, you can just imagine for example of sphere, and mentally change their color, shape or so on.

- If it is hard to imagine, you can simply assign a phrase or a certain form, such as a sphere. And also to nominate the changes that must occur.

- In any case the management (control) will still reach its result.

- In your consciousness already has everything. You simply activate your consciousness using images to implement your goals: both through logical phase, and through your spiritual structures - Soul, Spirit.

- Intention influences literally each cell of a human body. Intention creates emotion – an energy wave, an impulse or force which puts intention in action.

- Imagination - a management structure of the system.

- Humanity will evolve in the direction of absolute positive creation, expanding areoles of residence. (This is more like spiritual areolas - when a man sets himself tasks and solves them.)

The development of mankind can proceed as by settling of other planets – at first within ours solar systems, then in distant Space. (The Earth – a mankind cradle; but not eternally to be in a cradle – it is necessary to go «to learn the World» - to master distant Space).

Anyway the mankind should develop the consciousness to be able to supervise and operate surrounding reality (to keep fertile the earth, to create oxygen, plants, and animals and so on). – Here each person should aspire to come nearer in the development to level of God.

Thoughts, words and actions of man will be the main element, and machinery, buildings, planets, all of these and other physical structures will be a secondary element. And it will be the next level of life. People after development of this new system of knowledge will operate the worlds. It will be a completely different level of being. There will be no place to decay. This will be a constant renewal of the Worlds in which the status of Eternity is transferred to the next status of Eternity.

METHODS OF MANAGEMENT HEALTH. A REALITY

SPACE OF CONTROL.

Control space is in front of us at arm's length. In this space are created the main optical system. By creating an optical system - the sphere, it is necessary to put information into it. Information can be light, color, shape, sound, image, letters, and numbers - all this relates to the field of information.

There are two basic levels of management.

The level of control logic, the way we are accustomed to thinking the reality around us. Concentration on numbers refers to the level of logical control.

The level of spiritual control. At this level you can go when already mastered the basic course of major logic control technologies. State of spiritual control - a state of peace, joy and love. Here it is in this state we are shall always be.

THE BEST STATE IN WHICH TO DO SPIRITUAL MANAGEMENT:

- Must be the intention, purpose.

- There must be a volitional impulse, confident that now is necessarily succeed.

- The state of unconditional love for man.

- The state of stillness of mind and peace of heart and soul.

- The state of calm, inner joy, inspiration and sensory perception of the environment.

• A soft voice.

• Do not apply pressure - work smoothly

The goal of management should clearly and distinctively marked. Grigori Grabovoi not recommended to enter a state of trance. To make management must be able to clear the head, controlling the environment. Work soul and take your time. Do not doubt all you get. If distracted, go back and continue concentration.

First, according to the teachings of Grigori Grabovoi through your consciousness perform action to save everyone and to prevent a possible global catastrophe. Through this action more quickly to solve your personal problems.

And then the light of consciousness will spread to the light image of any event of the world and accomplish constructive management of any event.

Then go to the solution of personal problems: after setting a personal goal (talked for, write or think over) it is desirable to visualize a positive outcome.

ENTRY INTO ACTION OF MANAGEMENT.

Before control is desirable to talk over, write or think of the formula: (optional)

- "I act, how does the Creator"
- "In the collaboration with the soul and will of the Creator".
- "According to the norm of the Lord"

"The work will go on the goal of Creator to prevent a possible global catastrophe to the infinite creative and harmonious development of the micro and macro- salvation. The work is proceeding at the same time with my full recovery at the cellular level by the canons of the creator."

GOAL OF CONTROL.

- The absolute norm of health.

FORMULATION OF OBJECTIVES

- Properly formulated goal leads to success.

- Try to avoid the details in the description of the desired goal, generalize. You can use neutral wording.

- First, a well thought out purpose of your control.

- Ensure correct use of verbs. Use the present or past tense.

- On the negatives: avoid the prefix "no" because consciousness takes everything literally.

- We work only on the creation, use in the formulations positive terminology.

- Determine the approximate date of performance of his desire: the most rapid and harmonious for all management by objectives Creator.

- Put to manage your final wish (aim).

GOOD DEFINITIONS:

- The absolute norm of health (my or my family)

- Normalization of today's headlines

- A harmonious solution to the problem (housing, money, etc.)

- My financial well-being

- The maximum harmony for me and others improve my financial situation.

- Normalization of my personal life

- The normalization of relations with a particular person: wife, husband, son, daughter, friend, colleague, etc.

- Safe journey

- Normalization of the situation

Put a clear goal: “floats” goal - “floats” the rest of the information.

OUT OF WORK.

At the end of the work it is desirable to thank the Creator, or Grigori Grabovoi or stakeholders for their help. As to “close” control.

Fix results:

- Secure results in all parameters of reality.
- NORM Creator rested in all respects. Norm of Creator takes place in all parameters.
- The result of the Creator willed. So be it!

RECOVERY AND TREATMENT BY POWER OF WORD.

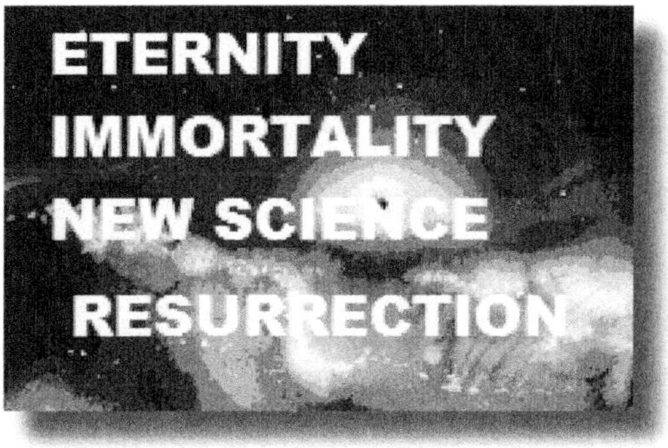

"The word - is one of the manifestations of the Soul."

To materialize the idea you have to pronounce it. The word creates vibration of space. Word is the technology of materialization.

Each letter, a word, a phrase, the idea has a certain vibration. The word can be written on a piece of paper and look at it. You can imagine it in a silver-white light. You can utter or sing.

You can affect the information of your consciousness with affirmations. Affirmations - this inspiring statement, one of the most reliable ways of mind programming, which can be recommended to all who are starting to work on the development of the consciousness.

Affirmations - words, thoughts and statements that say something positive and affirming. These statements are working as personal mantras and affect the mind through constant repetition and reminders. Repetition of affirmations can be replaced by a self-destructive and self-deprecating installation of consciousness on the positive and life-affirming.

A negative, dirty, nasty word destroys the DNA and makes a hole in the aura of a man. Negative words are like arrows, like weapons. We must have a strong light body to be resistant the negative impact. If you are in a state of peace and love, all the arrows fly away from you.

Affirmations can be repeated to himself, spoken out loud, write, and sing. Any sentence that has the positive sense can be used as an affirmation. Make yourself for bouncy, cheerful, short affirmations and repeat them as often as possible.

Repeating the proposed phrase, you will notice how your mood has improved.

With affirmations you are able to manage the information of your consciousness, and through it influence on other objects.

Examples of positive words and affirmations:

"LOVE", "IMMORTALITY", " I LIVE FOREVER"

"ETERNITY", "HARMONY"

"SUCCESS", "HAPPINESS", "FINANCE",

"RESURRECTION", "I AM LIGHT!"

"RECOVERY", " IMMUNITY", "I AM ETERNAL!"

"SAVIOR", "I AM LOVE!", "NORM OF MY HEALTH."

"INFINITY", "I ALWAYS YOUNG", "I AM IMMORTAL."

"I SAVE THE WORLD", "I AM LUCKI IN ANY SITUATION."

"NORM", "PROSPERITY.", "I AM THE CREATOR OF MY REALITY!"

"I matured EVER" "I am SO WANT AND SO WILL", "I - FORGIVENESS"

"I AM ALWAYS IN A STATE OF CALM, EMOTIONALLY BALANCED"

"I AM CREATED IN THE IMAGE AND LIKENESS OF CREATOR"

I personally thought up to my affirmation and sometimes repeat it. It sounds like this:

"I AM LIGHT I AM LOVE. Every cell in my body and all my body fill with light of eternity. All to what I touch fills by light of Eternity. My physical body is 18 years. I am beautiful and happy. My life is filled with success and prosperity".

In all these words, phrases and sentences laid impetus to the movement, to the desire to create as the Creator works! With this momentum people instantly goes to the very highest of vibrations and how to turn to the Creator listens to all by nature return to the vibrations of the Creator.
You can imagine the Earth recovery,
Create an outline of her Eternity and say the following affirmations:

«Saving the world, I'm saving myself"
«I developed spiritual abilities, I'm in harmony"
«I love myself; respect all people, transforms nature"
«Me action aimed at the creation."
«I am creator of my health, mood, and behavior.'

« On earth are born high spiritual children."
« Collective consciousness directed to creation, peace, harmony"
« I want so and so will forever."

If we focus on the past, we bring that past into the present, which in turn creates the future. We are the only ones responsible for what we are experiencing. If we do not like the current situation, we are the only ones who can change it. We are the creator in physical form, expressing ourselves in the words of matter. The only way to change the situation for the better is to face everything and avoid nothing.
We have Divine powers within us that have been forgotten and dormant.

Enhance immunity by the word «IMMUNITY»
Each sound is energy, the divine vibration. Depending on what word we utter and from its inner content depends on the energy we produce and send into space, man, animal, nature, the earth, the universe.
The word can raise, kill, and support the person to raise it, and so on. The word can be treated. A sick person can be treated with the word "immunity".

Imagine the word lilac-purple color and mentally slide it across the body: Immunity word many times as a helix surround the body clockwise. You will see a protective cocoon.
Word «HARMONY» is balancing the hormonal organs in the body.
Visualize the word voluminous letters silvery-white or lilac-purple. Enter through the crown of the head and consistently draw it through all the hormonal glands, one after another.

Word «NORM» has a multi-level understanding of the word and is unique in terms of impact on the controlled object. Possibility of words is limitless, but we need to understand the responsibility for its use in the management of the world and themselves.
Design of word for management is an important semantic function, acting on the transformation of the structure of the World. Building a phrase with a distorted sense bears the stamp of the distortion of the world, which focuses on all the elements and leads to a distortion in a state of physical matter and events in the world.
Working with the words "Savior," "Salvation," "Harmony," "Eternity", "Infinity" is the most powerful words that can be treated.

SIMPLE GEOMETRIC SHAPES OF INFORMATION

"Based on the principles of transfer of information in the forms, you can create any particular control over technological events, healing, including any disease".

(Grigori Grabovoi)

You can convert the information of the disease into health information, using simple geometric shapes: sphere, cylinder, cone, or a box. Anyone can easily imagine geometric shapes in their perception, and working with them is simple and effective.

Sphere is an element, which is an absolute coordinated system. Or absolutely harmonious form.

Example:
The disease can be represented as a geometric form of a selected color. All the properties that the disease has, can be placed in this form and can be "sent" away, gradually turning it into a point, which will disappear into a distance.

At first, visualize an action in your mind about the rescue of all and of prevention of all possible global catastrophes. Through this action your personal problems will be solved fast: then take care of your personal problems. You can either let your purpose speak for itself, or you can write it down, or say it in your mind. It is desirable to see (mentally and sensually) a positive result.

Activating Macro-goal of salvation, you get the help from the Creator in the solving of your local problem. You are acting for yourself, as a Creator, as you remember about the problem of Macro-salvation for the whole World.

It is very easy to work with a sphere as an object of information.

A sphere has a single coordinate, radius.

- By changing the radius of a sphere you can easily shrink it to a point or stretch it to infinity.

- A sphere can be seen far away, as it seems to be condensed, and close when it is increased in scale. If necessary, it could even be considered by layers.

- A sphere can be increased or decreased according to your wishes. Its main features are: the outer shell - the task of Macro-salvation. Inside - are your goals and objectives.

A SPIRITUAL ASPECT OF HUMAN

Color is a property most suited to the spiritual aspect of a person due to the fact that a color is perceived as an infinite value. By working with color, people actually act on infinite sets of information links. Languages Color is a reliable tool for human self-knowledge. Depending on colors, the environment is changing our behavior, condition, our health, and mood.

The effects on the body with a particular color, i.e. the corresponding frequency vibration influences in a certain way on the organs in the body.

Enumeration of different colors can be treated to recover their strength to help others. You will notice how you change, or how the person you are helping is changing. Get to the level of sensations and feel that every situation has its own taste, smell, and color. Work in a positive and your positive vibrations will increase.

1. Iterate over the front of your mental colors of the rainbow. When your eyes catch hold of some color, perhaps the particular color you need at the moment, concentrate on it for 5 minutes, pass this color through you, and see how you feel. Apply this color in clothes. This method is very effective.

2. Build small spheres, each of them filled with one of the colors and learn how to move them: zoom out, zoom in, and push to the side. Merge spheres into one, watching the colors change, shuffle them up as if juggling them, string them onto the unit.

With these manipulations, listen to your body, how sick and healthy organs react to color in their movement. Hold this feeling in mind: Recorded in the memory color will restore the required events in the future.

KNOWLEDGE TRANSFER WITHIN A SPHERE

If you want to transfer knowledge of Teaching to a person, you can do so as follows.

In the person there are points of reception of information.

1. "Charge" the sphere with information of teachings of G. Grabovoi "Eternal and Harmonious Development".

2. Mentally put a sphere with teachings to the face, chest, and into a right hand of a man. Make a circular motion around the body. Information from a sphere gets retracted into a body.

WHAT IS CLOSE AND DISTANT FRACTION OF CONSCIOUSNESS

Close fraction of consciousness is when an exact fraction of consciousness is within our grasp of knowledge (not abstract), it is information that we already possess about any subject, action, situation, and so on. This is what we see, hear, and feel around us, just like other information that we have receive from different sources (hear, say, TV, and so on).

The distant fraction of consciousness is information that can only be received by us, it is the information, which we can get from other sources, and this includes abstract knowledge.

For example, the knowledge of any star from a constellation like Alpha Centauri is the distant fraction of our consciousness. Basically, it is the information that is just very abstract. We are simply guessing what it's about....

The consciousness has a light golden color to it.

«**Super far areas of consciousness** provide a super fast processing of the information. Therefore, when you use the farther areas of consciousness, you can make your controlled events and recovery in very effective and fast way by processing of the information».

A FEW EXAMPLES OF TECHNOLOGIES OF CONSCIOUSNESS

The task of the teaching of Grigori Grabovoi is a multi-dimensional optical perception of the surrounding reality, through the structuring of consciousness, the structuring of a soul, so learn to manage the environmental reality, as does the Creator. It is very important to learn how to work with streams of light (beams). In the words of Grigori Grabovoi: "The path to the light is the way to God". You must be able to perceive and recall light streams. This is necessary in many controlling technologies, as in the process of control there should be lighted optics, lighted numbers on which we focus. Light stream - is our inner light; it is the ability to see the light of inner vision that connects us to the Creator.

When working on technologies of Grigori Grabovoi, there are two main stream of light that can connect us to the Creator. It's bright white, with a touch of silver glow - the perception of a very concentrated light. In his lectures, Gregory pays great attention to the luminous flux of Eternity. The luminous flux of Eternity is bright white, very intense light substance, which has a golden shade. Saturating the space around you with a light, the structure of a soul, consciousness, we create light body.

If it is difficult for you to make or imagine something, create a mental picture of an object you want to have. Tell to yourself that the given object is what you would like to receive or create.

For example:

1. So, (if you have difficulty with the above), you can tell yourself: "sphere is created".
2. You have created a sphere and want it to contain all the information needed to cure this disease. It is sufficient to say that this sphere is already contained.

3. If it is difficult to find an effective method of healing, you say that the disease is gone; the healing has begun in the best way and at optimum speed.

Activating Macro-goal salvation, you get the help of the Creator in the solving of your local problem. You are acting for yourself, as Creator, as you remember about the problem of Macro-salvation for the whole World.

MANAGEMENT BY THE SPHERE:

1. Visualize near left shoulder a small sphere of silvery-white color. Assign that sphere to be the Universal Salvation. Macro-salvation.
2. Near the right shoulder visualize the sphere of any other light color (gold). Put into this sphere your chosen goal.
3. In your perception, merge both into one. This is at the front of the chest.

At the moment of connection you will see bright light, as silvery-white. All done.

MANAGEMENT BY TWO STREAM OF LIGHT

Concentrate and imagine that to the left side from you is the silvery-white column of light, beaming up. Mentally identify the column of light with the Teachings of “Salvation and Harmonious Development”.

Imagine on the right side of the light another column of any other bright color (golden, purple, pink), and mentally put into this stream of light clearly defined task or a specific problem. Indicate the dates and options for solving the problem.

Concentrate and visualize the two column of light stretching up to infinity, where they merge and form a common stream that contains both colors that comes at you from above.

“Try to look beyond where the light is, which comes from above, as if the combined hue. That color has to fill the space around you. Try on deciding your task, to solve the problem of others.

YOUR ETERNAL DEVELOPMENT

Aim yourself or concrete another person onto a path of Eternal Development. Let this person (or you) go down the road towards Infinity (to the eternal development), where the image of that person (or you) in the future is corresponding to his (your) Eternal level of Development.

So by going down this path - the information of a person begins to change, and all of his diseases, and in general all the unnecessary information, all the problems – slowly, as if as a thread, unwind and disband, disappear on the way of this motion.

You should see this thread in your imagination. And now accelerate this process by simply pulling the tip of the thread. Keep pulling out this thread, realizing that it will take away the disease.

As a result, the disease disappears and goes fast enough to replace diseased tissue with healthy tissue. Gradually, but fast enough. Thus, it is possible, for example, to regenerate a missing lung.

DECIDING ON ETERNAL DEVELOPMENT AND REJUVENATION

Now is a new time, new era, when it is no longer necessary for a person to leave his physical body. The person can make the decision that he absolutely accepts his physical body and can develop with it further.

But thus, this person signs a new contract with God that he will live in this physical body, rejuvenate it, and thanks to this physical body the person is going to solve a considerable quantity of tasks. Then the body of the person will start to rejuvenate; there will be a regeneration of a whole body.

Therefore clear properly built in terms of target control your way which is fully under control of yours and which is completely stable because it has initially existed what truth of the world which has cone originally with you and beforehand with God deeds and then deeds of entire world on the whole. This way is non changeable and you may develop and live calmly eternally.

REJUVENATION

Those who believe in Salvation, in Eternal life from the perspective of

Macro-salvation Technologies, must be young - they rejuvenate very quickly.

There is a point of view, such a position in which it is believed that young or old - is a certain age at which a person should do this or that action.

And in the collective consciousness of these phases (young or old) - it is as certain paths and some frames that capture a kind of average normalized system of perception. This means that there is a need to change the collective consciousness. And if you decide not to age, change your mind. Do not take a conventional dogma.

"What is the age for the human mind? - It is needed to react."

If your conscience allows it, you can reduce your age at any time, whenever you want. It all depends on how you have adapted into the environment.

REGENERATION

Regeneration is a process of a complete recovery. Everyone should be healthy, young, and happy. There is the True Reality - the World of Eternal Youth and Beauty. Human beings who engaged in rejuvenation, are able to initiate the True Reality through themselves. The question is: What baggage of "collective consciousness" you are focused on in your personal consciousness?

COLLECTIVE CONSCIOUSNESS

Collective consciousness is the unified consciousness of all people.

There exist stable views of everything. Every man has his own ideas about everything, differing from the point of view of other people. The resulting averaged view or conception - is a collective representation. And that it is perceived by people as an objective reality.

There are two realities:

1. The one that exists in our minds;
2. Second reality that is outside of the mind, what we perceive as something stable.

All elements of the world are created by the joint consciousness of all living people. If at least one part of the consciousness changes, the world will also begin to change, transform.
Consciousness affects matter on the physical reality. Thus, matter is dependent on consciousness. Therefore, based on the collective consciousness, it can change the Earth, space, and other worlds.

ERASING VIRUS DEATH IN THE COLLECTIVE CONSCIOUSNESS

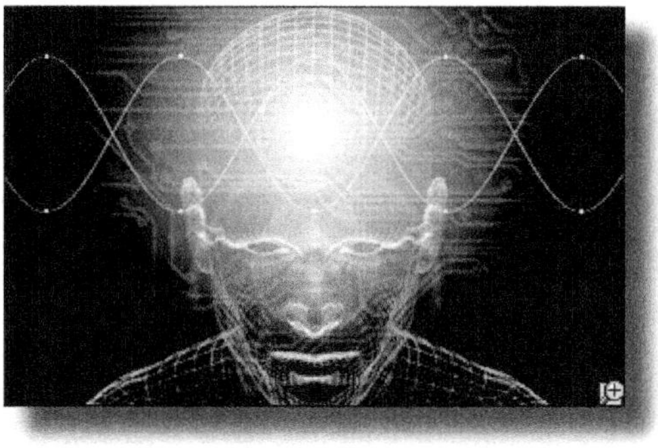

1. Place a sphere of the "Collective Consciousness" in front of the zone of management (in front of chest)
2. Visualize within that sphere a virus of death. It has a dark color - black or brown.
3. Visualize that the virus is lightening up, it is twisting, compressing in to a point and finally disappears.

MAINTAINING YOUR PHYSICAL BODY THROUGH GOD'S LOVE

The physical human body has the ability to obtain all the necessary energy from realm of existence (ether) - which is synonymous with the Holy Spirit.

God's love is the best way to enable the human body to become and remain healthy. Realize the state of Universal Love for the world and feel the love of God inside you.

Now, allow Love of God as a living energy into your physical body:

You can breathe in through the pores in your skin the pure Energy of Love. Here we have to visualize how the energy of the Essence of Love of God out of the ether (the Holy Spirit) as it permeates through the pores in your body and spreads all over your body, penetrating into every cell of the your body and nourishing it and rejuvenating it.

<u>"REPLACING THE INFORMATION IN THE COLLECTIVE CONSCIOUSNESS"</u>

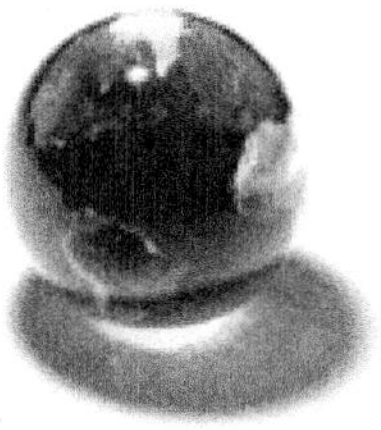

The man has a task out of the influence of the collective consciousness. It is necessary to overcome the speed of recording by our minds of the concept of life, which has implications on the genetic level. We need to overcome the rate of incorporation of biological decomposition (decay) of the body.

System of effects is tough and powerful in the collective consciousness. The person should change their level of consciousness to be able to withdraw from the collective consciousness.

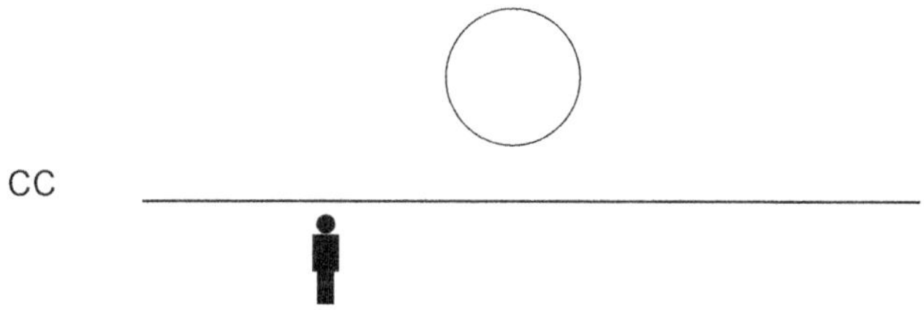

We must acquire a higher vibration, to be "eased" through the expansion of consciousness.

Inside **COP** a man is just an element. Through the expansion of his consciousness, he comes out of the COP.

To do this, rewrite the formation of consciousness to "eternal structure".

THE TECHNOLOGY OF "NORMALIZATION OF LIFE EVENTS"

Imagine a beam of light ahead of you, running into the Light of the Creator. This ray – is a "norm of life events". See: a ray of light near by "event norm" for a particular person, it can be rough, discrete, or of varying degrees of illumination.

Direct the sphere of the "norm" to a particular person on the ray. It rolls along, leaving behind a glowing corridor. Somewhere it must push through a force of will and bring to the light of the Creator and to take a backward wave. You can do it every day. It creates your life and destiny.

A

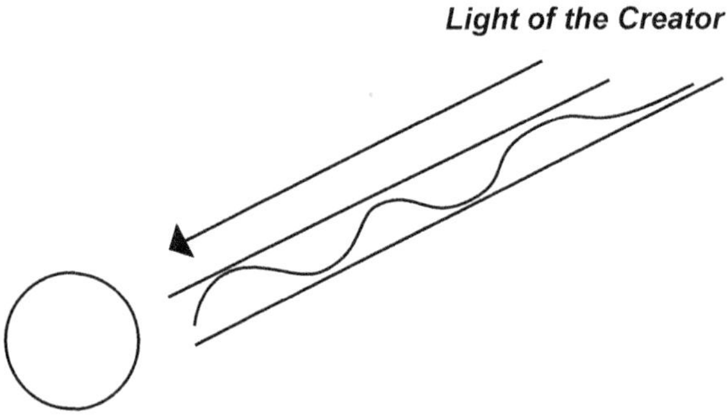

1. Prolongation in the future.

To consolidate the elements of progress, we must copy the result and put it into the future. It should be directed to the infinitely stretching line of your sphere with the result of work that is done. Use this sphere to send impulse to the Light of the Creator.

B

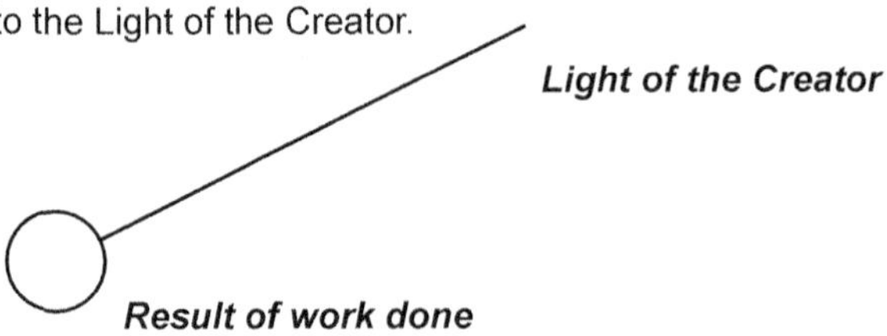

Or: create the sphere of a future and sphere of a present. Impose same way to the sphere of future silver-white color flowed into the current (sphere).

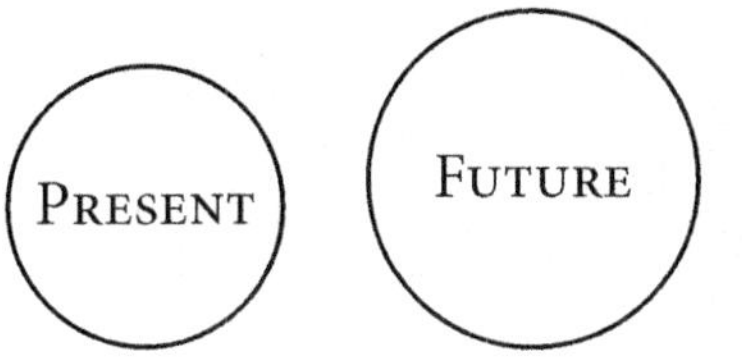

HEALING NUMBER CODES

"Every disease is a deviation from the norm. Deviation from the norm in individual cells or organs, or a whole organism. Recovery from illness means a return to normal... Behind every number is the corresponding vibrational structure. The same can be said about the sequence of numbers... and focus on the sequence of digits indicates tuning. Likewise, the sound of a tuning fork tuned musical instruments"

(Grigori Grabovoi)

Every problem has a numerical sequence (code). Each situation can be expressed by numbers. The ability to digitize the problem comes in the process.

Grigori Grabovoi has published two important books on this subject: "Healing Number Codes" and "Methods of Concentration for Every Day".

Every number has a spiritual and energetic structural vibration. Numbers are not just mathematical signs, they are the energy of the Creator. You can cure diseases with specific sets of numbers.

A similar structural vibration also lies behind every word and every auditory sound. The same applies to a series of numbers, as each number vibrates within a definite frequency. Series of numbers have their own correlating structural vibration.

"Number Codes" help to structure our consciousness levels which is the platform for control of our health and events. Be open and receptive to this unique creative process. The goal is always the NORM, and the NORM is the initial, unchanged or undamaged spiritual state.

Grabovoi's method of "Healing Number Codes" to cure diseases is as simple as it is effective. In his book he names several diseases and assigns each one a specific and unique Number Code. These Codes may well consist of five, seven, eight or nine digits. Ten, eleven or twelve digits exist for the control of events. Focusing on a specific Number Code helps you heal yourself of this disease. Numerical series have an impact on the body even without concentrating on it. But by concentrating on these numbers, the impact is greatly enhanced.

CONCENTRATION WITH NUMBERS OF SERIES FOR EACH DAY FOR EVERYDAY CONSCIOUSNESS DEVELOPMENT

These exercises will help you to develop health, consciousness, wealthy events in life, and harmonize your life with pulse of Universe.

I suggest accomplish these exercises everyday. For everyday of month, we provide three exercises for management the events. Basis of exercises is concentration on the goal, which will be achieving. You can choice a goal of concentration, for example, healing of illness, world understanding, etc. Mainly, you must regulate the information for everybody's salvation and harmonic development. You accomplish world salvation job, and; therefore, this regulation can realize the information war against disaster.

Practically, concentration will accomplish in following actions:

1. You define goal of concentration in some geometric shape, for example, sphere. That sphere manifests the goal of concentration.
2. You spiritually prepare to build the necessary events as God realizes that.
3. You control sphere location during the concentration on the objects, and move that sphere in most enlightened area of perception via will effort.

That is only one of possible variants of concentration. You can find many other. Very effective concentrations based on genuine world understanding.

In first exercise, you will concentrate on element of outer or inner reality; in second exercise, on sequence of digits; and in third, on word form of concentration.

Efficiency of concentration depends of your relation to it. Try to use creative approach, and hear your inner voice suggestions.

You can write digital series on sheet, and then concentrate on it. Another way of concentration is following.

Concentrating on nine digits series, you can imagine that you and goal of concentration as some ball locate in sphere center, and the digits had been writing on inner surface of sphere. You must find the most enlightened digit, and; after getting the thoughts about it, you must connect that digit with inner ball.

Concentrating on seven digits series, you can imagine the digits on some facet of cube, and you can move these digits according to your feeling for succeed maximum effectiveness.

Let describe another way of concentration. You can connect each digit with some element of inner or outer reality. For example, you can connect one digit with some feeling; and another, with some tree. That is up to you. Thus, you manifest the digits with reality elements or your consciousness elements.

These examples give you some management possibilities. You can change the concentration structure, mood of it, and symbolic manifestation of digits. As result, you will achieve more efficiency, and better time management, which is very important in everyday life. When requires fast salvation, your concentration will provide it.

If your goal is harmonic development, then time is not so important. Base factor is here harmony (in any circumstances), which will be providing by your concentration.

We assume individual character of these exercises. Choice of your development system cannot accomplish based on logistic level. Surely, you form your goal, you intent to realize it, but goal already existed in your soul, and that goal formed earlier. Therefore, concentrating on some goal, firstly, will realize goals of your soul, which are goals not only of your development, but also the development of society.

Accomplishing these goals, you will feel (on the soul level, on the level of Creator) that this is exactly what your must realize in first step.

Thus, when we are talking about concentration, we, firstly, are discussing the world harmony. Need to understand, the harmony always contains as necessary element the salvation, if a situation requires it.

Main goal of harmonic development is creation such conditions, in which never appear the dangers. We must care about eternal harmonic development.

Created and approbated concentrations in each day of the month, I dedicated for that goal. Accomplishing these concentrations, you will develop your path of development as continual and joyful, you will salvage yourself and other, and you will live eternally.

Based on these concentrations, you can actively influent on any situation. Your understanding of concentrations action as world salvation and harmonic development will open freedom, which Creator had given to you. That forms creative development of everybody and your happiness.

We provide concentrations for 31 days. If you accomplish concentrations in February, which contains 28 days, then after concentrations for 28th day, you must follow to concentrations for March 1. That means day must always correspond with day number in concentrations. Any time of day and night is acceptable to concentrations. For each day, you will define number and duration of concentrations. We suggest concentrations before important event and systematically.

If first exercise is too complicate for you, then you can remove it, and accomplish other two. In any case, you will get the results. Along the exercises, you will understand the first exercise deeply. Thus, you can accomplish the concentrations, which you like and understand.

1 DAY OF MONTH

1. You concentrate of right foot. That is basic point connected you with outer world. You are basing in thoughts on Earth. In your consciousness, Earth is basis. Your basic point of foot is simultaneously the point of maintenance and point of creation. You can develop your consciousness based on point as point of creation. You recognize the principle of development the plants and your physical body. According to that principle, you are able to create any outer reality. That understanding is basis of concentration.

Nevertheless, you not need to think about that deep mechanism. You can only concentrate on your right foot, and imagine event, which meets your goal of concentration. Mechanism of concentration will work automatically, and you will get the event in harmonic way, because that concentration provides harmonic development.

You can execute that concentration many times in the day.

2. Seven digits for concentration are following 1845421.

Nine digits for concentration are following 845132489.

3. You concentrate on world, on all objects of world. You feel any object of world is part of your. Perceiving that, you will feel slight wind from every

object, which gives you the solution. When you will get that every object has your consciousness particle, you will watch Creator's harmony.

2 DAY OF MONTH

You concentrate on little finger of right hand, and simultaneously hold in consciousness the desirable event. You can concentrate many times in day with different duration. You must relay on your intuition and inner voice as in other concentrations.

During this concentration, you can do something and touch some subject with little finger. It not interferes.

Important is concentration on little finger, not on the other. Management will stay harmonic.

2. Seven digits for concentration are following 1853125.
Nine digits for concentration are following 849995120.

3. You must consider world harmony in connection with you, and you must create that world as God creates. Watch on the world and you will see the picture from past. Watch on the world and you will see the picture from present. Watch on the world and you will see who you are. That world is forever.

3 DAY OF MONTH

1. You can concentrate on the real existing plant, and you can watch it along the concentration. Another approach, you imagine some plant, and then you concentrate on its imagination.

Concentrating on plant, you form necessary event in light, which reflected from the plant. You construct that event and really see it. That method called method of reflection. Event, which constructed in that concentration, will be harmonizing, because the plant exists in harmony of world.

2. Seven digits for concentration are following 5142587.
Nine digits for concentration are following 421954321.

3. Watch the reality and you will see many worlds. Watch your world, come to it, and expand it. Watch world as witness. Near to world, and put your hands on it. You will feel warm of your world. Move it to your and watch Creator. Hear what He says and what He suggests. Compare your knowledge with that and get the eternal world.

4 DAY OF MONTH

1. You concentrate on crystals or stones. Concentrating, for example, on

stone, you imagine around it some sphere of information. You watch in sphere all necessary events, and; therefore, you place those in sphere. In such manner, you control the events.

2. Seven digits for concentration are following 5194726.
Nine digits for concentration are following 715043769.

3. Watch the reality from perspective, which induce harmonic methods. One method continues another. When you are going via street, you are feeling current step is appearing from previous. If you are sitting, then you can stand up. Every move is different. Move roots in previous move, and create next. Consider world as continual, as world in which every move touch you only. Monolithic world gives you the concrete methods of management. Come to that world, take it in your hands, and your hands will be the world, which hold your world. You will see that you touch the eternal world – world of all worlds. That world will be unique for everybody; world, which you choose and everybody choose. Create world, which is ideal for you and everybody. Ideal world is monolithic. You must watch everybody's perfection and your perfection in your monolithic world, which is monolithic for everybody.

5 DAY OF MONTH

1. You concentrate on reality elements, which appear from your interactions with other reality elements. Let us explain what that means.

Considering some subject, you concentrate your consciousness on that subject. Therefore, that subject got some part of your consciousness and knowledge, and transferred those to other reality elements. Similarly, sun light falls on subject and then reflected light falls on other subjects.

Thus, if you concentrate on some subject, you interact with it, and then that subject sends somewhat to surroundings. Your goal is to think and illuminate what was being sending by subject. You can concentrate on that, and then imagine the necessary event. In this method, you accomplish concentration on manifested secondary element, which created the subject of your concentration.

Well, based on direct vision, or logical considerations, you find what subject has transferred to surroundings (after interaction with you). Concentrating on that corollary – secondary element of concentration – you imagine desirable event, and realize it.

2. Seven digits for concentration are following 1084321.
Nine digits for concentration are following 194321054.

3. When you watch sky, you know that Earth existed. When you watch Earth, you can think about sky. If you locate under ground, then sky exists above it. This truth must be a source of eternal world. Connect Earth with

sky, and you will see that everything what existed under Earth can be above it. Go to your spirit, and find resurrected people where they are. Move the infinity to world truth and you will see the eternity of world, and genuine Creator. Because Creator gives to you what you have. Because you creates as Creator do. He is close to you. He is your friend. He loves you. You give Him your hands, and create as He has created. You are His creation, and you are creator. Only Creator is able to create the creators. You must be in harmony with Creator. You must be open for Him, and must be eternal in every manifestation and appearance. Everything, which you want to correct, can be correct ever. Everything, which you want to create, you can create in your location in any time. Eternity exists for perfection. Creator's actions fill the Eternity. You are the one who watched by Creator. You are the one who want Creator realization in Eternity. You watch yourself in Eternity. Creator, which presented in you, acts with you in every your actions. Call to Him, and you will have a harmony.

6 DAY OF MONTH

1. You change density of consciousness structure via perception the remote objects. That method is convenient for realization the desirable event in proper place. You need to concentrate your consciousness in that place.

In contrary, your can use that method to prevent some non-wealthy event or situation in proper place. You must dissolve the negative information through disperse consciousness in that place. Dispersion of negative information leads to event dispersion.

Realization of desirable event accomplishes via consciousness concentration, which based on remote elements of your consciousness. We already discussed that method. Here you will use consciousness elements, which responsible of perception the remote objects. You can watch real physical remote objects, or objects in your imagination. In both cases, you will use the remote elements of your consciousness. If you simultaneously imagine the desirable event in proper place, then this event will realize.

Well, more remote area of your consciousness with information of event provides faster and more complete realization of event in proper place.

In relation to destructive forces, you can use dispersion method, which disperses negative information and prevents the undesirable event.

2. Seven digits for concentration are following 1954837.

Nine digits for concentration are following 194321099.

3. If you watch upended world, you must realize that upended, compressed, or dispersed world is always world of harmony, wealthy, and unity. Creator's wealth lays in basis of such states of world. You can get that harmony from knowledge about your eternity, and any structure and

information will not be able to change that God's Will.

7 DAY OF MONTH

1. You concentrate on super-remote consciousness areas. In practice, we deal with it when we watch the remote clouds, or far objects as trees of those leaves. You must process lots of information for materialization some object or realization some event. Super-remote areas of consciousness provide super-fast processing of information. More remote area provides faster processing.

Knowing that, you watch a cloud with physical vision, or in thoughts, and simultaneously you build desirable event on that cloud (you can use any remote object for that practice). The desirable event will realize in fastest way.

Event realization will accomplish in harmonic way, because cloud cannot destroy and damage somebody.

2. Seven digits for concentration are following 1485321.

Nine digits for concentration are following 991843288.

3. You see, that world develops in image and status of your actions, which accomplish in correspondence with God's will. You see that everybody recognizes world. If you are going to change it, then keep up a correspondence of your actions with everybody's wealth. Consequently, your action will realize, your health will improve, and everybody's wealth will come. Everybody's wealth is deal of world, which leads to God's Realm, to eternal life for everybody.

8 DAY OF MONTH

1. This day you learn to manage based on corollaries of event.

Imagine, you are sitting near to some lake, and watching a moving launch. Water is still, but behind the launch appear waves. Waves are corollaries of launch move.

Watch a leave on tree. We can consider the leave as corollary of tree existence.

Clouds cover the sky. Shower begins. Each drop we can consider as corollary of cloud existence.

Many such examples exist around us. You take some event and concentrate on one of corollaries of it. Simultaneously, you keep in mind desirable event. That event will realize.

This method is very effective. Based on it, you can change the events from the past.

2. Seven digits for concentration are following 1543218.

Nine digits for concentration are following 984301267.

3. Infinity of digit contour of number 8 connected the worlds, which already considered in previous seven days. When your world will connect with all worlds, you will see that your soul is joyful, as worlds count. Perceiving each particle of world as everybody's joy, you will see that joy is eternal as eternal is wealth. In that status of joy, you will rise your hand up, and will see impulse of God's wealth, which calls to Eternity. Watch Eternity in place where it exists. Watch Eternity in place where it not exists. Watch Eternity in place where it exists ever, and you will stay Eternity creator in place, where it missing from another person viewpoint. When you will watch and create Eternity, you will stay eternal forever, and in any world. You are creator by image, and Eternity creates you by image. Creating eternal, you will create yourself. Creating yourself, you will create eternal in same manner as Eternity creates another Eternity, and in same manner as Creator created all of us simultaneously.

9 DAY OF MONTH

1. Concentrating, you transfer maximum remote areas of consciousness in nearest areas of consciousness. Your perception of remote and nearest areas of consciousness must remain the same for both types of areas. In that case, you can get united impulse for creation any world element. When you will succeed that, you will stay a specialist in management, which can keep everything in good conditions based only on desire and stayed in some spiritual state.

United impulse develops some spiritual state. This state not connects with mind, because thoughts in that state can miss. That state can have simple mood on good, creation, and harmony.

That mood creates wealth development of events. Based on that method of concentration, you illuminate especial form of perception. Perception locates in your consciousness, presents part of your consciousness, and you intentionally structure it for management.

Method involves deep knowledge of management, which based on consciousness.

2. Seven digits for concentration are following 1843210.

Nine digits for concentration are following 918921452.

3. Watching the world as core of world structure, you will see that everything (plant, human being, animal, molecule, etc.) has one God's basis. God showed mechanism of everything creation. Watching how God created, you will create too. Understanding that via beginning and deepness of your ego, you will see the ego development, and ego transformation in world. You are that world. You are the reality. Watch on that via world eyes, via eyes of everyone, via your eyes and you will see that your soul is your eyes. Watch world via soul and you will see world as it is. You can correct it, if that is necessary. You will see world as world

of Eternity. You will know your path always when you watch world from inside and outside.

10 DAY OF MONTH

1. You concentrate on the objects, which you perceive in one impulse of perception. First, you prepare for perception all objects in one impulse. Then, you must recognize all these objects in one impulse.

In the beginning of that practice, you will get partial perception of objects. Do not worry about that. Your target is maximum full perception of the objects. You will get this perception after some time of practice.

However, even in the beginning you will get some information about each object. For example, you will get the information about objects existence.

Getting the information about object, you must find point of concentration and prepare to impulse. Then, you can consider any object. You can get access to all management spheres. If you will perceive many objects simultaneously, then you will manage big information area.

As concrete example, I give you some result of that practice. Assume you watch a computer. One glance only will enough for you to understand how manage that computer, and what you can get from that computer.

That concentration helps to get information from any object, because you learn to manage information objects. Access to information accomplishes on logical or direct spiritual basis. Well, I had given to you concentrations for first ten days of month. The other concentrations you can find based on cause-effect connections in information area. What you already know, you can develop further based on fundamental management position. However, I will continue my description, but in brief topics.

2. Seven digits for concentration are following 1854312.

Nine digits for concentration are following 894153210.

3. Union of two digits, one and zero, gives you an ability to watch world from the beginning, when zero presents in one. You act when you increase one with zero in ten. You action must be harmonic. You must understand that every your action can qualitative and quantitative increase your manifestation. You are world manifestation. Harmonize it with what you are watching. Watch yourself and your thoughts. You must be in place where you are presenting. You must be in place where you are not presenting. Your harmony must lead to Eternity. Resurrection is Eternity element. Immortality is also Eternity element. You must find for yourself genuine Eternity, in which immortality and resurrection are just particular cases of Eternity. You must be creator of everything. You must clear imagine and recognize what is follow up after resurrection and genuine immortality. Genuine immortality creates next Eternity status, next world status, and next person status. You must recognize that other problems,

problems of Eternity, which you consider, create new worlds, which you build in consciousness. That world you will get when you succeed the Eternity. You eternity includes in you (like one and zero gives ten). You just need to understand that. Transfer to that level via logical action (as union action of one and zero), and you will get the Eternity in every your action, in every your manifestation, and in every your step.

11 DAY OF MONTH

1. You concentrate on interaction between animals and human being. For example, in your home live dog, cat, or parrot. Think about your interaction with them on deep level. What is sense of these contacts and interactions from your and they perspectives?
Understanding the processes of thinking and perception of all participants of interaction will give the possibility to enter in structure of reality management.

2. Seven digits for concentration are following 1852348.
Nine digits for concentration are following 561432001.

3. Number 11 is world manifestation (you add one to one), which is inside of you and which everybody can see. You are a being, which everybody can see and can get your harmonic experience of development. Share your experience with other and you will get Eternity.

12 DAY OF MONTH

1. You concentrate on events, in which assembled some object. For example, a goose lost his feather. In that case, you concentrate on action of restoration that feather. How you can restore that subject? You try to understand how to restore the subject.
Another example, a leave fails from some tree. How get back that leave on its place on tree?
That is concentration on assembly the separates objects in one, providing their normal state. Practice in that concentration develops management.
In this concentration, as in many others, you can consider yourself as object of concentration. You can restore any of your organs. Some women asked me once to restore her womb, which removed in surgery. Using this concentration, I had restored her womb (see document in Appendix B).

2. Seven digits for concentration are following 1854321.
Nine digits for concentration are following 485321489.

3. Unite with world in its covering. Unite with world, which you perceive it in your actions. You will see that your actions are core of world, which harmonizes with you everywhere and in any time. You will see, that

Creator send to you His wealth, and want to join with you. You must have union in place of God's development. Union with Creator exists in development. In divine, genuine, and constructive development will be accomplishing that union. You develop in direction to Eternity, and that will be joining with Creator. Life eternity is genuine union with Creator.

13 DAY OF MONTH

1. You concentrate on discrete separated elements of some object of reality. Let assume, you watch some object. That can be truck, palm tree, or stone. No matter what the object is. You consciously consider parts of object. For example, you can imagine truck, which assembled from many parts.

You can consider in such manner any object, excluding a human being. We must consider a human being as one object. This is law. Any other object you can consider separately. Your goal of consideration is detection existing connections between parts. You are finding these connections and simultaneously keeping in consciousness a desirable event, for example, healing somebody or getting the ability of direct vision. On this way, you will realize the management of that event, and will improve your management ability.

2. Seven digits for concentration are following 1538448.
Nine digits for concentration are following 154321915.

3. You will watch the faces, which created world before you. You will watch the mechanisms, which created world before you. You will watch world, which was before you. You will feel that you exist ever. Transfer that feeling to these faces, to these mechanisms. You will see that everything around you (created artificially or naturally) is Creator. He manifested you in what you are watching. Your manifestation is world, which will create. In such manner, you can detect any constructive technology of intellectual, spiritual, or technological development. Consider the development as equal development of any reality element and any information object, then you will see core of your soul, personality, and Creator. Individuality of Creator and everybody creation is base of world harmony, which existed always, in everything, and understand everybody. Creator creates everybody individually, and in one moment. Thus, you can create world individually and simultaneously for everybody, for any space-time.

14 DAY OF MONTH

1. You concentrate on moving objects. You are watching those and asking the questions. "Why the cloud moves?" "Why is raining?" "Why the birds fly?" "Why happens something?" You try to find information sense of each event.

If you are concentrating in such manner, and simultaneously keeping in consciousness the necessary event, then you will realize it. Correspondently, you will improve your skills in management.

2. Seven digits for concentration are following 5831421.
Nine digits for concentration are following 999888776.

3. You must watch your hands as hands reflected the life light. You must watch your fingers as fingers reflected light of hands. Watch your body as body enlightened with clear light of Creator – light of love, good, and health for everybody, and light of my doctrine about the eternal life. In that day, you can perceive that Doctrine of Eternal Life, my doctrine, and you can ask me in thought about your problem. You can ask me also in any another day and ask me about help in development eternal life and creation for everybody. Ask me and you will get help. You can ask yourself, and can recognize what you got from me. You can recognize that knowledge, and share it with others. In that day, you can harmonize with me (as in any other day). You can ask me also when the time will be not measuring with space-time. You can ask me about help, conversation, some event, and just to attention. You are free as you was ever. Take that rule, and expand it for everybody, then you will get eternal life in place where I am. You will get eternal life in place where you are. You will get eternal life in place, where everything exists. That principle will be genuine and true, as it is now. You are a person, who exists in Eternity, and you are Eternity.

15 DAY OF MONTH

1. In second day, you concentrated on little finger of right hand. Today you can use for concentration any another part of body (another finger, or nail, or any another part). Then concentration accomplishes similarly to concentration in second day.

2. Seven digits for concentration are following 7788001.
Nine digits for concentration are following 532145891.

3. In fifteenth day, you can feel Wealth of God, which sent from Universe Mind created by God. That Mind thanks God for His creation and status, which provides Him an ability to reproduce Universe. You can feel wealth of plant or animal in relation to you, wealth and love from another human being. You will see that you love them. Creation, Wealth, and Understanding are in Love. Love of everybody to everyone is Creator, which manifests world in your appearance. You are appearance of Creator's Love, because He is love related to you. You got present of God from the beginning, and you are creator, because Creator creates you. Go to Him. He is everywhere. Go to His call. His call is everywhere.

He is where you are. He is everywhere, where you are. You are in God's moving, and you are His Eternity manifestation. Go to care of Creator. He created eternal world in everybody's development. You will see that world is eternal, and manifests your Eternity. You are creator, which creates eternal. Creator created you as eternal when He created world.

16 DAY OF MONTH

1. You concentrate on elements of outer reality, which contact with your body. Are you remember sentence, "Sun, air, and water are our best friends"? You need to recognize contact with such "best friends."

You feel the warm and touching of sunlight.

You feel wind, sometimes quite power, sometimes very week. You can feel the non-moving air, or hot and humidity of weather.

You can feel the freshness of water when you swim, or take shower.

You can accomplish these concentrations in any season and you will recognize the interaction with nature forces.

You can practice these concentration everyday. If you will keep the necessary event during the practice, then that event will realize.

2. Seven digits for concentration are following 1843212.

Nine digits for concentration are following 123567091.

3. Feel harmony in everywhere. Harmony belongs to Creator. Feel the harmony, where it is and where it will be, where it was, and where it was not. That is harmony of transformation in eternal life. Go to yourself, and feel the harmony everywhere. You will see the waves of joy and happiness, which harmony creates. You will see that you create world as stabile and harmonic forever. You are warrior for eternal life and eternal belief in God's Divine.

17 DAY OF MONTH

1. You concentrate of the elements of outer reality, which surround you always. There are surrounding space, Sun, Moon, star constellations etc. You concentrate on some such element and simultaneously keep in consciousness necessary event for realization.

2. Seven digits for concentration are following 1045421.

Nine digits for concentration are following 891000111.

3. Watch the resurrection of everybody. You will see that world restoration is reality in which you live. You will feel that you locate in eternal world. Move forward along the path, and you will see Creator, which is eternal. You will enjoy your eternity. That joy is eternity of life and Creator. That is Creator, which created you. His Love is infinite. His simplicity is credential.

He is simple and clear as you imagine. He is creative and kind as you had thought. He is Creator of you and He shows your path. Go on that path. That path is your path.

18 DAY OF MONTH

1. You concentrate on static objects (building, table, tree, etc.). Choose somewhat of those. You need to detect individual core and sense of selected object. Sense of object is sense for you. You must understand what means that object for you. During the concentration, you must keep in consciousness the necessary event, which will realize.

2. Seven digits for concentration are following 1844212.
Nine digits for concentration are following 185321945.

3. You move in direction to people. You move in direction to events. You work in direction with resistance. When you watch that, the resistance will decrease, power of it will be weaker, and you will watch world of Eternity, even the resistance still there. Go and present everywhere. You are able to be everywhere. You can embrace the world of Wealth. Thus, fight the resistance for eternal life wealth, and the resistance will destroy. Then, you will see and accept the light of eternal life, and that will realize for all the time forever.

19 DAY OF MONTH

1. You concentrate on set of separated elements of outer reality, which appeared as result of one element transformation. Let give some examples. Cloud transforms in the raindrops, or leaves separated from tree. During the concentration, you try to find the laws, which prevent such process development. Core of that concentration is search of such laws. During the concentration, you must keep in consciousness the necessary event, which will realize.

2. Seven digits for concentration are following 1254312.
Nine digits for concentration are following 158431985.

3. Spirit fighting for your place in world, as fighting of your soul for manifesting Creator, leads to your control about intellect and mind. Your consciousness will stay universal. You will stay what you are. Your Eternity manifests in your considerations and your considerations stay Eternity. Your thoughts make world eternal. You will be there, where you are locating, and you will be there, where you are not locating. You will be eternal, even world constructed from time intervals. Interval of world will stay world if you will be there. Space will unit with eternity, if you will be there. Time will disappear, and you will be in move, in eternal time. You

will feel eternal time, and that time will come to you. Every moment of your time is eternal. Feel Eternity in every moment, and you will see that you already got Eternity.

During the concentration, you must keep in consciousness the necessary event, which will realize.

20 DAY OF MONTH

1. You concentrate on remote areas of consciousness. Your goal is help for other people. Imagine you need to explain something to some person. You will explain something, which that person does not know and understand. We already know that every person have full knowledge in his/her soul. Thus, you must help that person to recognize information, which he/she already owns. By the way, genuine understanding is recognition the information from soul.

Arousal of person to recognition information from soul we will realize via remote areas of consciousness.

Realizing that concentration, you involve in world salvation program. Principal thing of your concentration is management for everybody wealth, and positive development of events for everybody. No matter, is a person far or close to you, your management will help everybody.

We can call that management as concentration of everybody's success. We imply that according to your work, the events development will be positive for everybody.

You can add additional exercise in practice beginning.

You concentrate on remote objects such as Sun, planets, star constellations, etc. You not need to watch those via physical vision. Your goal is recognition the objects as objects of information.

2. Seven digits for concentration are following 1538416.

Nine digits for concentration are following 891543219.

3. Consider world from highest point of your consciousness, from deepest position of your soul, and with spiritual desire to everybody's wealth. Consider world as world, which created now, and create it as it is now. However, try to create world and change it state. Change and decrease the defects of world in direction to eternal life and positive development. You will see that the defects are not defects, those are non-correct world understanding. Recognize world correctly, as Creator gave it to you. You will see that creator is everywhere, and truth is everywhere. Just need to step in. Just need to not neglect, and come to truth forever. You will see that world changed. You will see that Universe is your, and Creator welcomes you. You will see that you are creator, and you can create everywhere forever. You are Creator's assistant and assistant for everybody. You create creator and you come to point of union for everybody. That union pint is your soul. Watch it and you will see life light.

Your soul creates that light. Light of your soul calls you to development. Light of your soul is world. You watch world because your soul watches. You watch soul, because you have soul eyes. Watch yourself, and you will see your union with world, which existed always. Your consideration is consideration of world. Your knowledge is knowledge of world. Disseminate your life knowledge and spread up your soul light and you will see eternal life in such state, in which you are. You will see that eternal life is long time with you. Eternal life is always existed. It exists. It will exist forever. Eternal life is you.

During the concentration, you must keep in consciousness the necessary event, which will realize.

21 DAY OF MONTH

1. You concentrate on digital series in reverse order. Let give an example: 16, 15, 14, 13, 12, 11, and 10. Numbers in series must be between 1 and 31. Thus, you have 31 numbers. Relay on your intuition when you create the series.

2. Seven digits for concentration are following 8153517.
Nine digits for concentration are following 589148542.

3. Watch on mountain brook. Watch snow melting. Imagine these pictures, if you watch those. You will see that your thoughts not distinct from your eyes. You will see that your consciousness not distinct from your body. You will see how your soul builds your body. Do not forget that knowledge. Transferring those from second to second, from you to another person, and creating from moment the Eternity, you will always build yourself as you live earlier without efforts. That eternal creation is eternal life. Build around you other objects, based on the same principle. Build the worlds. Create joy, seed the wheat, create bread, and construct the tool and machine. Do the machine non-destructive. You will see that in machine manifested God and consciousness. Stop the machine if it is danger. Build the body if it is seek. Realize resurrection, if somebody has gone. Prevent death of somebody. You are creator then act with harmony of world. Act with harmony of everything, which created. Act with harmony of everything, which will create in infinity and manifestation of world. Act with harmony of yourself.

During the concentration, you must keep in consciousness the necessary event, which will realize.

22 DAY OF MONTH

1. You concentrate on reality elements, which infinitely reproduce. For example, you can concentrate on Eternity conception, or infinite space conception.

During the concentration, you must keep in consciousness the necessary event, which will realize.

2. Seven digits for concentration are following 8153485.
Nine digits for concentration are following 198516789.

3. God creates your soul structure. God restores your soul structure. Watch how God creates and restores your soul. Your soul is in action of restoration. Open your world and watch where God restored. Watch mechanism of restoration and your will see Love. Love brings the light for world. Love is basis of world. Love exists forever and from beginning. Who created Love? Watch and you will see yourself. Your Love is you belonged to Love. Create with love, wealth, and great joy of eternal life and everybody's happiness, then you will see everybody's joy of your surroundings. Watch joy of everybody and happiness fills your heart. Be in happiness, be in harmony and that happiness brings to you Eternity. Watch via your eyes, watch via your eternal body, watch via eternal glance on your relatives, and give them Eternity. Watch via your Eternity on world, on your surroundings and give them Eternity. World will bloom and flower appears, which bloom eternally. That flower will be your world, which is world for everybody. You will live and your happiness will be infinite.
During the concentration, you must keep in consciousness the necessary event, which will realize.

23 DAY OF MONTH

1. You concentrate on development all elements of reality in direction of God's goal realization.

2. Seven digits for concentration are following 18154574.
Nine digits for concentration are following 581974321.

3. Watch on world. Watch on business, which must done. Watch your feelings. Watch the connections of those with events. Why you watch forward? Why you feel? Why your business develops in such manner? Word "else" cannot be presenting in world because world in unitary and different in its unitary. Why word "unique" means diversity? Feel nature of events in your concrete business. Watch your business from all perspectives. Watch your organism and restore it in conscious moment. Watch your consciousness and make it for solution of all your problems. Watch your soul and see that everything already is there.
During the concentration, you must keep in consciousness the necessary event, which will realize.

24 DAY OF MONTH

1. You concentrate on shape of human being and transform it in some another object (for example, videocassette, pen, or plant). You must see an element of human being, which can transform in videocassette, for example. That means - we need to consider a human being for transform his/her shape in videocassette.

2. Seven digits for concentration are following 5184325.
Nine digits for concentration are following 189543210.

3. You watch your reality. You come to reality, which is you. Consider all days from first until 24th, and you will see that your love is infinite. Watch world with love. Watch how you construct your feeling. Consider the feeling as eternal creation and you will come to love as to Eternity. You come to it always and will stay with it forever. Creator is your God. Creator creates you as loved. You are God's creation and you love. Love is life, and life is Love. Manifest your love everywhere you will appear. Manifest your love in places, which define and predefine you. You cannot express love in words, and you cannot express love in feelings. Your actions are love on the place, where you create.
During the concentration, you must keep in consciousness the necessary event, which will realize.

25 DAY OF MONTH

1. You concentrate on set of objects, which you choose. You classify objects in groups based on some property. For example, you can set in one group tape and audiocassette, because those accomplish one function. Tape and radio-device you can put in one group because those created on electronic base. In one group, you can put objects of one kind, for example, two different books. If you consider the book content, then you can put these books in different groups. You are free to choose classification property.
Staying in your home, you can use the surrounding objects for concentration.

2. Seven digits for concentration are following 1890000.
Nine digits for concentration are following 012459999.

3. Come to the thought about you. Catch the thought about you as your reflection. Watch yourself as you are watching the others. Watch yourself as you are watching everybody. Watch yourself as you are watching a tree branch, plant leave, and dew on grass or show on sill. You will see eternal things. You will see you are eternal.
During the concentration, you must keep in consciousness the necessary

event, which will realize.

26 DAY OF MONTH

1. You concentrate simultaneously on part of object and on whole object. Consider an example. You watch the cowherd and simultaneously some of cows. You concentrate on some cow and imagine cow's further development. Same way you can consider the ants and some ant separately.

Your goal is to embrace in one glance the whole and part simultaneously. Concentrating in such manner, you will get such ability to see whole and its part simultaneously.

2. Seven digits for concentration are following 1584321.
Nine digits for concentration are following 485617891.

3. Understand that you will develop eternally. Watch that your development is permanent. Involve in process, which develop eternally. Every move is eternal, and every thing is reflection of Eternity. Every individual is Eternity. Every soul is many eternities. Go to multiform eternities from one unitary Eternity. You will see that Eternity is one only for everybody. Come via that to your soul understanding, and you will see that you are creator of what you need. Apply that to creation of anything, and you will see that you created everything. Apply that to your organism creation, and you will see that you can restore organism in any time. Apply that to health of the others, and you will get the experience if you healed somebody. Everything restoration is always an experience for you. Create more kindness, joy, and happiness and you will get the Eternity as technological instrument of your consciousness. Expand consciousness on firm conditions of Eternity. Outrun the Eternity in infinity, and consider yourself as Creator's manifestation. You create on place, where the Eternity expanded. You are creator of Eternity, you control the Eternity, and you compel the Eternity.

During the concentration, you must keep in consciousness the necessary event, which will realize.

27 DAY OF MONTH

1. Your concentration repeats concentration of ninth day. Just add to it infinite development of every element of concentration.

2. Seven digits for concentration are following 1854342.
Nine digits for concentration are following 185431201.

3. Help to those, who need help. Help to those, who not need help. Help yourself if you need help. Help yourself if you not need help. Consider the

word "help" is expanded sense. Consider kindness as help manifestation. You are kind and you help. You are creator and you have help. Every action of your creator brings help to you. Everything, which you created, is help for you. You have an infinite number of assistants as you help to infinite number of the others. Lead society in common connections and help one to another to the wealth. Give happiness to everybody, and you will see yourself in harmony of world. World in which God-Creator is everything created around you and created by you. World is manifestation of God in surroundings of you. Manifestation of God as your creator will express in your soul as genuine world understanding, which self developed after accepting infinite life. Infinity of life is infinity of Creator. Being infinite living you must be infinite creating and co-creating. We not need to do something for being infinite creating. We created forever for being co-creating. You are able to do that every your thought, every your move, and every your action will create Eternity.

During the concentration, you must keep in consciousness the necessary event, which will realize.

28 DAY OF MONTH

1. Your concentration repeats concentration of eighth day with one important addition.

You, probably, remarked, we summarized the numbers of previous day 2 and 7: 2+7=9. In current day situation is different; we must multiply numbers 2 and 8. Thus, eight will double, and program of eighth day will repeat. You not need to repeat that concentration exactly; you need change it. For example, you can change somewhat in yourself, or in concentration conception. Using old schema of concentration, you can discover new aspect of it.

Your understanding and perception of that concentration must permanently expand and recess. Concentration is a creative process, which maintains your development.

2. Seven digits for concentration are following 1844512.

Nine digits for concentration are following 195814210.

3. Watch yourself as you are watching on world. Watch Creator, as Creator is watching you and get from that an understanding about Creator's expectation of you. Watch His look, and you will see it. You will see that Creator's look directed to far events of world too. Thus, your goal is management of these events of world. You must transform any event in harmonic event. That is your genuine goal from beginning. Creator already created and realized and your goal is follow that path, because you are "image of God." Creator had self-created, and created you too. Create yourself and the others. Create the others and give them wealth, then you will get world, which created for you, for others, and for Creator.

Create for Creator, because He created you. Create for Creator, because He created everything. Thus, everything created by you, you always created for Creator.

During the concentration, you must keep in consciousness the necessary event, which will realize.

29 DAY OF MONTH

1. You accomplish generalized concentration. Revise all concentrations from first day until 28th. You must accept those in one impulse; all concentrations you perceive in one moment of perception.

Analyze your work and create new platform for next month.

You can imagine all concentrations as some sphere and place it on some infinite line, in which you include next month as beginning interval. You will create new platform for your infinite development.

2. Seven digits for concentration are following 1852142.

Nine digits for concentration are following 512942180.

3. Watch world via your eyes. Watch world via your feelings. Watch world via all your cells. Watch world via your organism and via all your manifestations. Watch world, watch yourself, and watch inside of you. Watch world with understanding of it surroundings of you. Watch reality, which gives you life. Watch reality, which gives Eternity. You will see that any perspective gives one reality, which creates life and Eternity. Creator created that reality, and eternal life. He sees you as you see. He sees you as you not see. He created you. He is God.

During the concentration, you must keep in consciousness the necessary event, which will realize.

30 DAY OF MONTH

1. You accomplish first concentration on created platform. That concentration is basis for next month.

Concentrate on world harmony. You must find, enjoy, and wondering about it. You are amazing about its perfection. You are amazing of world harmony as corollary of Creator perfection.

2. Seven digits for concentration are following 1852143.

Nine digits for concentration are following 185219351.

3. Principle of your concentration of previous days in that day can be basic. In February, number of days is 28 or 29. Thus, on thirtieth day, you will concentrate on first or second day of March. That transition shows eternal life cycle. Find the Eternity in all your previous harmonies. Find harmony in that simple example, because only after February we have

joining between numbers 30, 1, and 2. Joining of the numbers, which are distinct in nature, says us about universal nature of anything. Find universal nature in everything, in any information element. Find universal nature in hidden and visible manifestations. You will see, you will feel, you will recognize, and you will inspire.

During the concentration, you must keep in consciousness the necessary event, which will realize.

31 DAY OF MONTH

1. You concentrate on separated areas of manifestation. For example, you watch some tree, ground under it, air above and around it. All that components are the manifestations, which joined in your consciousness. You watch in those an eternal life reproduction. Life is eternal. Recognize that. Remember that when you are watching and feeling the surrounded world, and when you are dissolving in it. You will get the understanding of Truth: LIFE IS ETERNAL!

2. Seven digits for concentration are following 1532106.
Nine digits for concentration are following 185214321.

3. Concentrate in that day on yourself. You are healthy and everybody around you is healthy too. World is eternal. The events are creative. You see everything in positive light only. Everything around you is always positive.

During the concentration, you must keep in consciousness the necessary event, which will realize.

Notice to concentration

You must define number and duration of concentration by yourself. You define the most important result of your concentrations. If you aim the result for some date, then add that date to concentration.

Remember, these exercises are creative, and develop you. Basing on that concentration, you will develop spiritually. Thus, level of concentration will increase permanently and infinitely. You will detect that your life will begin to improve (that is result of your development). You will take your life management in your hands.

These exercises develop your consciousness, provide health, and harmonize your progress with Universe Pulse.

A FEW EXAMPLES OF TECHNOLOGIES OF CONSCIOUSNESS USING NUMERICAL CODS

NORMALIZATION OF PHYSICAL WEIGHT

1. Enter a digital number **4812412** into a small sphere.

2. Squeeze this sphere to a point, and enter it into the abdomen.

3. Enter a digital number **1823451** into another small sphere and then place it inside the pituitary gland.

4. Visualize a violet-purple sphere 2,5 meters in diameter.

5. Put your phantom in the center, along with its target in a digit.

6. Send the sphere to the Stream of Creator.

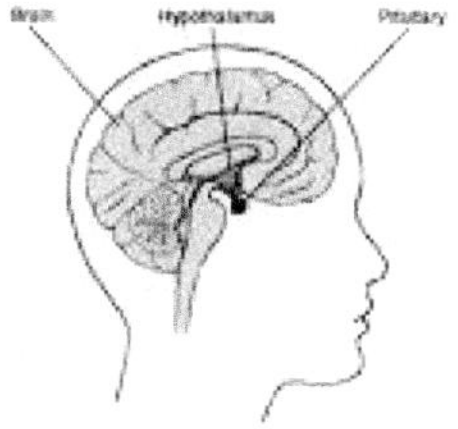

MANAGEMENT OF FREEDOM FOR ALL

Number of Freedom: **9189481**

Let's imagine a free world. All borders between countries have disappeared, all are free to move wherever they like. You do not need visas or passports. All people are peaceful and respectful of all cultures of other nationalities. Freedom!

PROTECTION WHEN YOU RIDE IN THE CAR OR OTHER

Preventing road accidents: **11179**

You can mentally pronounce this numerical code, or write on a piece of paper and put in the car.

TREATMENT OF SPINE
REMOVED PAIN REMOVAL OF PAIN

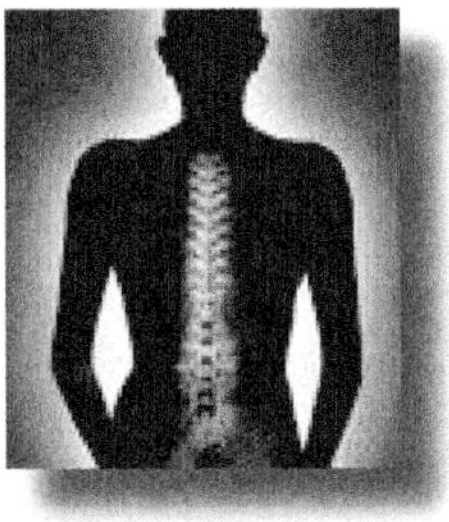

1. Enter **5481321** in the heart, or in the sphere and then into the heart. Heart – is the assemblage point.

2. **18548321** mentally lay out the figures around the neck stretching from one collar bone to another, like a necklace, so they light up. Keep it on for 24 hours or 2 months (from time to time re-activate it: mentally lights up).

And then - you create a sphere of 2 cm, and its connecting a bottom at a distance of 0.5 centimeter from the coccyx. And the task is: not to enter it inside the body, but to warm the entire spine. The perception of it can be felt like a warms of a sun. Expand the light, all the warmth of it targeted up.

WEALTH -MONEY

Normalization of the financial situation: **71427321893**
Surround yourself with this digital side, write down on a piece of paper and put it in a purse, imagine number in his office in his apartment, put the number in the document, or passport.

UNIVERSAL NUMBER OF SUCCESS

71042 - Number of an instant resolution of all issues for all occasions (good luck, success). Through this technology, you can stop or extend the time.

If you are late, create a rainbow from a place where you currently are to a place where you need to go to. See yourself there at the right time. Under the rainbow imagine a silver-white digital number **71042** without glancing at your watch.

ADDITIONAL NUMERICAL SEQUENCES

88888588888	Good health of all people
1814321	The absolute rate of health
19751	Instant recovery of health and harmonization of events
285555901	Harmonious family relationships
25122004	Resolving issues and problems
212309909	Stability of Business

881881881	The development of clairvoyance, extension of spiritual vision
71931	Protection from any adverse impact
19837198	Safety of a flight on an airplane
4748132148	Instantaneous stopping of the negative events

938179	Emergency Assistance
972319891	Absence of conflicts in the future and the current time
5217319	Make accurate decisions in a difficult situation
3175142179	Prevention of any accidents for 10 years

39718514	Providing eternal life
721498175	Complete freedom
972154218	Resurrection of all people
9738197	This numbers is introduced when there is not enough time
9371857195	Prevents possible conflicts and scandals. Takes into account the outdated problem
97318541218	Ecological safety anywhere
1258912	Critical conditions
4185412	Acute intoxication
1257814	Acute respiratory failure
7794218	Protection against viral infection
1001105010	To be at the point of balance. Inner peace
811120218	Working with plants
555142198110	Working with animals (to close sequence interval needs to be between 1 and 0)

In the book "*Recovery of Human's Organism by Concentrating on Numbers*", Grigori Grabovoi includes more than a thousand different diseases and medical conditions and provides for each a numerical sequences to cure them. His book consists of 27 chapters (groups of a different disease) each of them has a numerical code to cover related diseases of that group.

14899991257814	**Acute respiratory failure**
14899991895678	**Acute heart failure**
14899998915678	**Heart failure (clinical death)**
14899991895132	**Traumatic shock, shock and shock-like state**
14899991258912	**Critical states**

14899995813212	Sepsis
14899991289435	Circulatory diseases
14899998148888	Rheumatic diseases
1489999 5823214	Respiratory diseases
14899995321482	Digestive diseases
14899998941254	Kidney and urinary tract
14899991843214	Diseases of the blood system
14899991823451	Endocrine and metabolic diseases
14899994185481	Occupational diseases
14899994185412	Acute poisoning
14899995421427	Infectious diseases
14899991234895	Vitamin deficiency disease
148999918543218	Childhood diseases
14899991489145	Obstetrics, gynecological diseases
1489999148543293	Nervous Diseases
14899998345444	Mental illness
14899991456891	Sexual dysfunction
148999918584321	Skin and venereal diseases
148999918574321	Surgical diseases
14899991851432	Diseases of the ear, nose, and throat
14899991891014	Eye diseases
14899991488514	Diseases of the teeth and mouth
14899991884321	Unknown diseases and conditions

1489999	**Norm of laboratory results**
1884321	**Unknown diseases and conditions**

Human body consists of seven parts. To each part of the body applies a specific digital sequence. If an unknown disease or condition occurs, concentration can be practiced with numerical sequences related to that part of the body where the disease is located.

1819999	**Head**
18548321	**Neck**
1854322	**Right hand**
4851384	**Left hand**
5185213	**Trunk**
4812531	**Right leg**
485148291	**Left leg**

RESURRECTION OF GRAIG

I would also like to shortly address the topic about how the knowledge that Grigori Grabovoi has given to the world, the Technology of Consciousness, works. One of our dear companions, Lilya, has an acquaintance Barry Solomon, who one day called her in the middle of the night in desperate need to help his dying son. Soon our Boston's group joined to manage the norm of health for Graig.

We askedBarry Solomon to write in his own words about what occurred and the incredible outcome of the event. This story is truly a miracle.

Dear Lilya, I woud like to share with you and your group the horrible events that happens to my son.

My son Craig, who is a dialysis patient collapsed on the street in Cambridge at 11 pm on Saturday night due to dangerous levels of potassium in his body. When he fell his heart stopped. It took several minutes before the EMTS could revive him. When my son got to the hospital he went into cardiac arrest again. For the next 3 hours his heart kept stopping. He was given 16 electrical shocks to the chest to keep him alive. The doctors had him on life support. When I got to the hospital the doctors told me to say goodbye to my son. His body was ice cold, his face bleeding badly, and he was in a coma. I called Lilya at 4:00 am in the morning and asked her for help. At 6:00 am my son again went into heart failure. That morning the doctors gave my son dialysis to remove the potassium in his body, my son remained in a coma. That day he was given a brain scan and we were told that he had severe brain damage and if he lived he would never be the same.

I had Graig transferred to the Mass General Hospital where his doctors where that night, Craig was in a coma for the next to days. When he came out of the coma, he did not have any short term memory. My son spent almost one month at Mass General. There were many difficult days of physical therapy, severe pain from his wound+almost daily dialysis treatments. He also developed a severe case of pneumonia which was treated with large doses antibiotics.

The doctors call him a miracle. His doctors asked me if I believe in God because they coud not understand how he survived and how he had full brain function.

Graig still has wounds to heal and probably oral surgery on his teeth, but it is truly a miracle hi is alive. He has regained all his memory+has no side effect from the accident. He has even started to drive again and thanks everyone for there help.

I believe my wonderful friend Lilya helped save my son"s life. Lilya has a gift of healing that she gave to my son. Lilya thank you so much for this special gift you gave me.

I also want to thank the group of the healers for helping to heal my son. Between you, the doctors and the many people that prayed for my son health, I will always be grateful.

Barry Solomon

TEACHING ABOUT ETERNAL LIFE

Author's seminar held by Grigori P. Grabovoi in Russian in August 04, 2003

Good day one more time! The theme of today's seminar is "Grigori Grabovoi's Teaching of Eternal Life". In this theme I tell how to spread the technologies of Eternal Life on almost any line of events. Within this the events are supposed to be those that you do the governance either in the point of technologies of your health and in the point of, generally speaking, almost immediate access to any part of space -time. And also in such a way study the process that you look at the reality like at this one that at the same time is here and in another part. That is when you act here, you at the same time spread your governance on any part of space-time. So the meaning of Eternity means in this case as too personal character in the governance, because Infinite for a person is realized in a spirit aspect as an action of a person here and is spread for all informational objects.

And when I talked about performance that is the governance of your health, at first I meant that the meaning of health also includes the governance of line of events as eventual systems of events and structures of close reality.

And within this when we make the governance according to my "Teaching of Eternal Life", we study also a sphere of events, which forms almost every following step of life, every following element of life in the sum with practically all phenomena of life and also all phenomena in the point of eternal development of life. So this Teaching, this structure of Teaching is purpose governance, making its way to ensure the parameters exactly eternal life. So this is a real system of making outer governance, a real structure of interaction a person with a system practically governance both its personal tasks and the tasks, will say, of outer World.

So, at first, here I consider, in this structure of Teaching, consider a chapter of a personal exactly relationship to governance. That is you see the governance as your own aim, that is in this very level the main is to form your aim in such a way, that you have an access governance to every part of space-time owing to form your personal aim in interaction with the whole outer reality, which also exists eternally.

And in this connection the meaning of eternal life here first of all is considered as the life in the condition of development of any object of information to the side of Infinite. So, for example, such a simple, as it seems in the everyday understanding process, as treatment of disease, in this case is a process, directed to realizing of interaction of your personality with the whole exactly outer environment, and also inner system of your development. And in this way this reasoning must be considered in your governance that you interact like widening from you the whole space of the event. And in the same way, that is, you also widen the time, don't you? That is you encourage the time practically the eventual aspect in the governance.

And according to this, in my "Teaching about Eternal Life" I consider a macroplast, those one, which regards the level of saving everyone, as a level of development a time interval in action. It means that every element of time, personal time of a man, carries an element of infinite development. And with the point of spirituality of this phenomenon, for example, if we consider a Physical Body of the One God, here His meaning of time is squeezed till the level of space which exists out of time.

The Physical Body of the One God exists out of time, so it happens, that the governance of the time is an over high concentration of substance, for example a physical substance, spiritual. And in general, as an example, by the governance, it is possible, we may say, for being expressed, by governance of Soul giving property is characterized, that means there is such a condition in Soul, which is called exactly as governance of Soul - and this is an indication of Eternity. It means that any governance of Soul is an eternal characteristic.

So an applied aspect of exactly giving governance is consists of those, that you consider the governance as action of Eternity, such an action, which is directed to you. When you look on the logical phase of the Consciousness, that the giving action is the action, like being spread for all phenomena of the World, you see that if we consider there a remote object of reality, we suppose, a star or some remote event there, or the event, which you remember, that is that you put them in one line, you get an outer counter of your own body. Taking on purpose a various principle of events and spreading this principle on you, you see, that an outer counter of your body is what inseparably linked with those phenomena, which about, for example, you have put some, we suppose a system of governance, and this system is so that you see, that the phenomenon is too enough inseparably linked with you.

That means that the sense of giving governance is that you practically do not share yourself and the whole outer space. That means you do so that practically the whole outer space, it like grows from you; there some remote star it immediately appears here like on junction of your body and the star-itself. And when you begin to enter this spare of governance, you see that any line of events, for example which also have attitude to you, some diseases, this is those, what you can easy remove out of limits exactly such a close contact-with Eternity, we say. So if you have a contact -with Eternity very close, haven't you... with eternal development, that is, naturally, -with some element, which reduces the giving level, let's suppose, we have a deal with those, that we need to take out a virus or some disease from the organism, in this case it is necessary to enter vibration level of giving condition and any other phenomena are forced out. Because in the infinite development there cannot be a place for diseases, which cause in the organism, we suppose, emaciation or some other more problem conditions.

So the interaction of the system of exactly eternal development this is first of all also a technology, which acts at once you have begun to work. So with the point of this chapter of my "Teaching about Eternal Life" the governance begins at once since the moment yet before governing action. That is just it is possible to enter the spiritual resonance with the performance of giving level, and you get the governance straight away directed to eternal development.

When we consider like a structure bearing, that is a platform, on which Eternity is placed, the structure of Love: Love to the World, God's Love to the World, to the World's development, we see, that Eternity is a concentrated in fact Love, expressed like -with all elements of the reality. That is if all elements of the reality its aim of infinite development will like multiply by ideological purpose of eternal development of all at once, we get that Eternity is a concentration. Well that is just it looks some formally, but really, if we, suppose, pass on simple meanings there, such as weight, density, concentration, that is what we get. Logically it is very simple to consider in a logical plan. As soon as we pass on the spiritual scheme from the logical one, we see that Love is eternal too. That is we see the Infinite level, directed to Love to every person. That is The Creator God, the One God, He made himself in the way His expression of Love to every person is an infinite expression. He does not do in general once movements. That is the God always does those construction which exists eternally. Because He is out of time and in fact He can be, can't he... out of space, then He passes on a spiritual aspect of governance.

Consequently, a man's soul in mine, exactly, chapter "Teaching about Eternal Life" is a governance system, which is both out of space and out of time. That is we practically have the level, which characterizes a whole system like a system, well generally it is ideal, in the purpose aspect it is finished, in the purpose aspect for the giving person, why mostly personality is impersonated with Soul, but with the point of infinite development, the task of Soul is to do that the aim coincides with that how does the One God do. That is Eternal God, He has Physical Body and eternal development. So it comes out that to rule events according to giving chapter and in the giving structure of my Teaching, I show that the simplest elementary events of life there, which a man has - here, for example, he is, we suppose, in a usual system of there some optical connections, he considers something, does not he?... makes something and any element of this process must be that one, that every action directs a person and really gives him a technology of eternal development, eternal life.

Then we consider the process, well for example, clearly subjectively, let's suppose a man does some movements, as a person, he considers some processes of governance, nothing prevents him from being eternal. That is the man does not guess anyway that there he is like a essence, does he?... like a personality - this is a final phenomenon. That is in principle it does not lay intrinsically in a man.

And when you consider if it lays, we may suppose, in that, let's suppose, object there an animal, does not it?... dog, tree, it is evident, that there it does not lay too. Because the growing tree, the fact of the tree means that it is growing, there is no any information about the fact that now the tree is not growing. And in this aspect the present time is the feature of infinite manifestation according to the objects, which are expressed in this time. That is in this time we may only say that if there is a dry branch somewhere close, in the theory this tree may develop in the direction that a branch is dried. But this is an extrapolation of a human mind, based on some previous facts.

If we do so that to consider a position of the mind in such a way, that in the giving sphere we do not think, but do it in the sphere of physical body of the One God, it is so that this tree can be not destroyed and grow eternal. The same spreads to a man and any other object of information.

If we transfer such a technology of our mind on a bomb, as an example, do not we?... well, a bomb, which can be exploded, as only we put this technology of the mind in the sphere of functionality of this bomb and spread it on the eternal development, in the infinite development such an object of information, which must be exploded, but it may be not exploded, it means that not every bomb explodes. And when we like spread the giving moment on any aggressive source, the source of radiation, let's suppose, do not we?... the element of codification comes practically at once, as only you transfer Eternity in the structure. That is in the eternal development... What is an eternal development for the bomb? (tell, it is just metal, is not it?... with some supply of explosive. There in some thousands of the years it is restructured. Even within good conditions of storage this system is not functional with the point of explosion.

So if to place an aggressive object in the system of eternal development, we have a norm system. If we want, we suppose, to study a radioactive source, which there, doesn't it?... shines radiation somehow strong, with the point of eternal development it stops shining. By the way, I carried out tests specifically by myself, well, even not tests, but such practical actions on purification of planes after radiation influence, then, practically put a finish to a normal level, so, a source of radioactive emanation, laboratory source, only with a simple transference of the source in the structure of Eternity exactly. So we take the layer that is Eternity exactly, cut out like from the form of information, inserted in, and instantly the radiation reaches a background level. In this way it is possible to protect yourself with the point of what... or other people there, isn't it?... or to refine a technique, planes, seas and so on, territories, just to bring in the structure of eternal development any aggressive level.

And with the point of the God, just the One God, there is only a constructive development in Eternity. So and the constructive development is a norm, generally speaking the simplest logical norm of the whole development. As Eternity, in any case, I even gave the example with the bomb, we suppose, it becomes rusty, doesn't it? And so, when we consider well, we suppose, a technology of protection from terrorist actions with the point of a logical system of rendering harmless, then in this case, first of all it is necessary to consider projection on the layers of Eternity. That is how to do to prevent a terroristic action? It is enough to use the screen of Eternity, which is formed by your heart, and it is formed very simple: you take a rhythm of your heart and listen to it, and when you take it, you in fact do such a view, that you transfer this rhythm of heart in the luminescence on the level of your Soul, deep condition of your Soul exactly, where Eternity, made by the God Creator at the beginning. From the technological point of view it is very simple: just a sound of your heart, which you hear exactly with the inner ear, even not with a physical one, but like with an ear of Soul mostly, you can see, that there is a projection here, so a screen is created, which is likely exploited in the outer atmosphere very quickly, so a man begins to radiate the light.

With the point of exactly my chapter of "Teaching about Eternal Life" you can see that a man is a source of the Light, which is radiated from him, it is like a system of lenses. So as only you begin to see these displays of the Light, you project events on is light simply and see what is necessary to do to prevent a terroristic action, so, we suppose, to do some specific movements, that is to report some information or to act on the source, that is much better, making other levels of tasks for them, well, constructive, isn't it?... and the terroristic action is prevented. You know, it happens too often in the life that just changing of events, changing of the mind or a situation makes that negative event is not realized.

Another question is I that all people must work that no terroristic actions happen and not only those, which threaten the whole civilization. As if only one person stops to shine this light, such space and phenomenon are created, inside of which a negative situation can happen. Consequently, this is the task of compulsory teaching everyone be able to use the technologies of action, as the One God does, in the way like, generally speaking, a human body particularly develop intrinsically, doesn't it?... the whole world is the task of governance, generally speaking, of any event. So if we need, and we really need all events are favorable for a person, then it obtains that everyone must be taught.

And with the point of Eternity's technology, we almost have a very universal system of teaching. As an example a concrete case from my practice is a treatment of the fourth stage of cancer, when a part of a lung was not visualized, so it was just affected with metastases. With the point of replacement this part of the lung, well, nevertheless these are metastases, which are too strong volumetric mass, and when there was a talking about replacement of the giving mass, for example, by a normal substance, it cannot be cut out with a surgical way, because metastases were not only in the lung, but in the bronchial tubes and so on. And there is no other way to take it out, beside exactly the element, which is natural for a person.

This is a technology of replacement of such a large- almost a half of the lung was affected with metastases, I have produced, as an example used the following element of the governance. I just took an element from, well, that information: what is the eternal development of this very man, how does it look in the information? With the point of eternal development there is a transformation of structure of, well the following, the process, but slowly as a thread is uncoiled. And when you see the process it is easier to pull out a thread, than to cut a part of the lung, isn't it? And it is so, that it is enough to pull a tip of the thread just with the element of logical governance and the substitution happens immediately. So the replacement by a normal lung happened quickly there.

Within this I showed you that the point of entrance by treatment a disease exists doesn't it? You operate with the eternal process of development that is you consider it. When you consider it, you can see that the process of governance is not so large, as with the point of Eternity every cell is too small. And with the point of eternal development a person in general, the mass of his body is a very insignificant mass. So it is enough- in this case enough just to make a movement with a thought on the optical level, and a man is made a complete recovery. And this is so because I just stretched out the mass on the eternal time and on eternal development.

And It is received that, well, isn't it?... a simple technological system. But when we study how an outer World is made, as an example, don't we?... take a screen, and, which is likely directed on the outer World, and consider, that the outer World is also a system of, for example, some optic elements of governance, a very simple level of Physical Body of the One God and the body, which is eternal and can be removed in the form of human physical body sufficiently free there, then all forms body are what an interaction of events, which is directed on Eternity, means. It means that a form of the human body is an interaction of events, which is directed on Eternity.

So when you interact with the help of your form with the system of dynamic development of future processes or your ones, you get a very intensive system of governance, which you always have. The main is to identify, for example, the way you interact with all events, like your body, frankly speaking, governances events, if you, for example, do not do any intellectual efforts. Then you see from this form of Eternity that interaction always happens. And the combination of bodies, isn't it?... a lot of physical exactly human bodies occurs the next level of Eternity, more qualified level, where the following space and the next time is made.

And, for example, if to consider the situation, that many people are in a big hall, then if, we say, to isolate mass and substance of this hall, the more people are entered here, the more we can see the levels of creation, well, the initial creation of molecules likely, for example, oxygen, the primary creation of substance. It is so that extension, extension, for example, of a physical human body exactly, that is according to likeness the God's Body, we see, that the governance of substance with the point of substantial development happens.

So, such a concept that if limitation exists, so it is possible the problems with the quantity of people on small spaces may occur, really it does. not exist. The more people, the more elements for the next development are. And how, for example, it will happen? It is necessary though to adept urgent spaces for this, it means to adept with the point of logic and development, so with ... constructive development exactly.

And when we consider the process for the Infinite future, here there is exactly the next - the level, which is closely like a spurt in development- the next level, anyway, cannot be, it is obtained, realized only with some special technologies and only by individual persons, and still it depends according to the level of the previous level. So an individual person, however it may be, he still cannot pass away very far away. There is a story of information, which tells that it depends on the previous level, from where it had appeared, hadn't it? And then the next development happens nevertheless, when the first level is worked through. The conception of working off every element of event:, including a social event, what now I am talking about, exists in the Infinite technology.

It means normal life conditions, this is there the principle of Love in all life events, and this is the understanding of the way how the Physical Body of the One God does, how the World is settled down with the point of its consciousness as personality, just personality like a human. So when we consider all these elements together, we almost get an immediate governance of the norm. So everything that is going on we consider normal, but you know, this is that when normal takes place, then ... this is, you see, a governance too, so this is a correlation of different systems.

Then if to become lost in thought why it is going on normal? Because fixed influence of Eternity in the giving point is a level of joint governance, which you interpret like a norm. So it is necessary to be able to perceive the norm. And that's why with the point of Eternity to balance the governance, with the point of psychology and so on, you should understand the way you interpret the norm, that is to meet the norm. Here I introduce a special level of governance, which I have never talked about before, because every seminar I perform and create just in the moment of holding the seminar, so now I am talking for the first time that the level of contrary governance in respect of your governance exists. So you enter the point, like the vector point, observe where the optic of your governance functions, and create the contrary vector stream. That is that in fact you act in the one point from both sides, and at the same time your governance functions on you Without fail, so you insure to get like a reflected optic signal. But the One God, He made the World in this way. Every element of the reality, it reaches Him, He sees what had happened with it. To see, you ought to !mow and control, oughtn't you?... ought to have an effect in the movement directed on you.

So practically with the point of just my chapter of "Teaching about Eternal Life", we must see that our signal we had sent comes back almost at once. So the time of sending and getting is generally one point of information. "When we watch, this point of information is creation of primary impulse, these are we by ourselves. When we look father ahead from here, we see that to enter the structure of the physical body of your own from this point and to consider every cell, to consider in the permanent variant, Without haste, with the point of nobody forces us to do it quickly, we may consider well quickly, generally speaking, all correlations when the body. So when the body becomes clear and laconic with the point of our view, so we see one point, but understand everything around us, that is around - we just understand it. It is so that in technologies of Eternity there is a space with right consciousness of reality taking into account all correlations. As an example, here I consider: a plane is flying, well, for example, it may fall down. I study a reason- this is about one person. Within this I have to do it does not fall down. I, you know, consider the position at once at the same time, it means the space of comprehension. In the structure of Eternity the space of comprehension is likely exists or forms absolutely.. And according to my specific saving practices, we suppose multilinked systems, a plane is a multilinked system, which has a dispatcher, passenger, it is possible, for example, it consists of some problem devices for flying and so on.

And if it is necessary to ward off, as an example, terroristic action or to do some aggregate does not activate, or to do in such a way that, on the contrary, everything activates, then only one complex decision exists. So it is that this decision must have only one variant: saving a specific plane or satellite and so on, or just a person.

Then the one element of the giving governance is still only extrapolation, the situational development on the eternal development, because in this case we see a single side in conditions of combination of all events, as it just in mathematic language. When we consider all possible combinations of event, we get only one, which ought to be taken right now. Well, for example to call air traffic control officer, who gives a right command for crew team, don't he?... or some other decision, remote, to extend a level, the reserve level of events of some concrete person on the board. Then it acts in a way through reflected optic signals, and the plane, for example, arrives well. So in this case spreading to the eternal development exactly gives almost a single level in your actions.

This is that you ..... everything correct, as you respond to the eternal construction in this World. And when you want to understand how to chose, how to make extracts of this eternal construction, where it is placed, at first pay attention to yourself and your stereotype of thinking. And you will see that your consciousness is mainly about your physical body. Seldom enough a person is able to think, in principle, somewhere in another system with a strong-willed effort, there in some Galaxy, he imagines thinking near himself in fact. Of course, he may move out his thinking in some Galaxy there, but in spite of this it comes back. This return is some kind of periodical vibration of Eternity. Eternity - the farther we move away, the closer we move to ourselves. Just make a clearly practical experience and you will see that if you, for example, begin to hold away the thought very far, the pressure on your physical body occurs.

As some kind of governance, even rescue, come on, take and begin to push out the dangerous particle as an example, some isotope from some part of building, which is not needed, and begin to push out the information very far away, somewhere to the Universe. More you push out, more you see: luminescence comes on you. And it happens so that, generally speaking, the luminescence is harmless, but still you see it is shining on you, isn't it? So you have a superfluous channel of information, that is unnecessary source shines on you. And intrinsically on the whole only the God can look after you. So intrinsically as a matter of fact no one is permitted, if you do not want of course, do you?... someone looks at you when you are on the stage or so on. Only the God can looks after you and the rest •you are entitled to do that no one spies on you and control you and so on.

And when you, for example, do the governance of such a type, removing some microelement far away even harmful, shadowing occurs; you begin to disturb this principle. So a task is to solve the problem in the point of its existence. And for this it is necessary not to hold away, but to know how to direct your thinking in the way that the thinking transforms this point into the norm. There is a very simple way to do this, the other one: to direct all efforts of the outer World and your own ones on the norm. And with the point of technology of Eternity it looks very simple: take an outer system of the space, any removed star or just any Galaxy and install the optic rate in the point, which you rule and install it from you. If the point of the governance is some large specter of events, if you want you can extend it or use it in the view of one shining point and build a system like some corner, we say, where the events are in the corner. And the sense of the governance is that the governing event cannot be placed in the corner, as every informational element is free.

So if you are controlling a person, you cannot dictate any conditions, can you? Even if he is aggressive, he is ought to be given the conditions he could develop somewhere another system of events: to get out of aggression and find some elements, you see, it happened so that to create social conditions around him and so on, psychological, maybe even treat.

And when you do the event yet, the result is this comer optic combination is expressed in that the event must be turned around very free, so to be like both in outer side of the corner and inner one too. And when you begin to gather the corner, well, you see, you can press it, these two optic rates, the inner and outer Worlds, the World appears; and so in the projection you see your own body and the signals, which come from your body. In such a way you may study what level of the body to activate in according to what event.

As I told you the technology is very simple. You just work with yourself in the giving technology. You need not to influence or force anyone. It is enough for you to see the source of luminescence near the body and just to activate it, well, for example, a sphere near the knee of the right leg, well, for there is no there, for example, even some global catastrophe in outlook there, for example, in three years, you see.

You take this sphere, just look that it is so, for example, that it just must be filled with luminescence. Take and fill it with luminescence, simple, elementary personal luminescence. You are also a source of Eternity, a source of eternal luminescence, as the God primary created you eternal. Everything he creates is eternal. It is so that you do according to the God's will.

And when you just make giving governance eternally, an eternal quantity of times, in the logic you can consider the situation so, that you enter the information there it existed an eternal quantity of times through will effort. Really enter just once. Then it is so that you enter an act of Eternity in your action. So when you, for example, enter the act of Eternity through will effort, you do in the same way as the One God does, just situational, like technologically you do so. And here your will and the God's will coincide, they coincide so that the God made everything in the way of the eternal development, and you do the same.

Another question is there are such elements like temporary rain, isn't it?... foliage falls down there and so on, but for the One God every element of time is identical. The meaning of the next development is the same meaning for Him as those one, which was some when in the past, a hundred, thousand years ago and so on, isn't it?... milliard years ago. And it is so that with this point of view for the One God there is only a time of development, with this point of view how a person grasps time, the time of transference, the time of changing an object, a form of information.

And when with this point of view we consider the element of Eternity, here at first there must be a person, who governances with a form. A person can, for example, to exploit an open -cast, can plant a tree. Yet it is logically obtained that it is better for a person to be directed to the side of eternal development according to the God's tasks as he has more systems of governance. And this how the God made a person. As creative and action elements do not occur without will of the God. Then a man can create like he is, if he knows this system, which comes from the God. Then a person can create a similar one, knowing this system coming from God. Then he can reanimate, find where the knowledge of reanimation are in Soul, in what place of Soul. If you see your Soul from this depth of consciousness, well, we may say, he knows how to make eternal the other man. And when you begin to work in this technology, I gave such an example: as a hall there is, which, well, for example, is controlled with the point of view of physical aspects, the quantity, we suppose, of air, but when you come to those, that a man can create by himself, he can already show it always, everywhere and out of any hall. He can create the substance by himself, can materialize there, to create the substance, for example, to grow tissue, which are lost and so on ... And when you move to this system of governance, you see, that outer Light, coming to you, this is the Light of activity, the Light containing knowledge, the Light containing information and the Light containing in the point, in general, not only physical knowledge and in the essence not only physical knowledge, but the Light you see is like all manifestation Light of all events. So, all events exist in one element of Light.And when you, for example, want to recover eyesight, don't you?... or to recover the organism through eyesight, such a technology exists, you may just see the Light inside, in your Soul. So watch: and where is this Light in Soul? Yes, there is the Light which is grasped through eyesight outwardly, but you see this One in Soul. This is the same. As only you see this identity you will see just the God's luminescence, this is where the God's shines Himself.

When you see it, you see that this is Eternity, which is like multiplied on space and events. Does it mean now that I already carried Eternity out of limits of space and events. Like the God, He is out of space and events. It means that in the Image and Likeness any element of reality, including Eternity, can be placed out of space and events. But this is you have the most powerful in fact technology of governance of the whole reality, just carrying out the governance out of events, and for example, out of space and time, yes.

So we have a very simple essence: Soul is exactly in this place, there, where the Physical Body of the One God exists. Consequently, Soul exists in the same place, where the Physical Body of the One God exists, in that space and time. So this is another space and another time. Characteristics of this space are the essence of Eternity in every element of activity. Logically, it can be imagined as: if to take some movement and consider, that, suppose, the giving movement is the final one, for example, yes?... how with the point of Eternity to give him the functions of Eternity? The functions of Eternity are to spread in the meaning of ideological action in the giving case there. The One God performs that everything is eternal, and we, knowing that, do identically, don't we? Ideology is what moves people and moves with the point of eternal development. It is so that the ideology performed by us, performed aims are those, that they identify the structure of Eternity within the final actions. If a man does one step and this step is directed to the side of eternal development, it means that he gave functions to Eternity. So a man can give functions through his mind. The God gave him he made the space around him, the opportunity to think exactly as the One God tells and shows, and teaches, doesn't he?.. So when we make like extrapolation, try to understand that some or other real phenomenon of reality, for example, the space of Eternity, where the Physical Body of the One God exists, is corresponded with our one, we may say, that this is our development, our thinking, our future and we can governance it.

And when we consider the question of governance, for example, as absolute, that is, for example, everyone governances, and everything is good for everyone, that is those that in projection when a man is good in something particularly now, yes, any situation - this is a projection of this condition. Such information exists in future, that everyone is ought to govern with the point of norm of the event, they all develop eternally constructively, for example, happily, yes, everyone is healthy. This space shines, and when it is shining now - this is a norm of every individual person. Even logically, if there is no a common norm in future, opposite luminescence with the point of technology of Eternity does not exist.

With the point of technology of Eternity only the interaction of multilevel exists always, which has like a level of absolute norm, and your personal level of governance or behavior there. So when you enter some large supermarket, you need to find something among a great variety and to find quickly, and to get a positive event; or you have to be out of some situation, we suppose, problem, yes?.. to recover your health in general in situation of problem systems somewhere to get right decision -just go out the level of Soul. Such condition of Soul exists, where there is a norm only, and negative events do not occur there. This level forms through the way of luminescence existence, which is directed into Eternity and has its origin from the left part of body - somewhere near spleen, pancreas, and heart. Corner system of governance occurs. This system is for entering and meeting, as a satellite meets in the space, with a system where an absolute norm of the point of eternal development exists, well, according to my descriptions, such a shining space, where all processes are eternal.

Then you receive that this is the outer situation, it like works instead of you, that is it perfects the system of governance into the norm already because you have connected up to this in right way. You know, it is enough to have right knowledge, not necessary...or right way, isn't it?.. not necessary in general to know where this way is, well on what ground it is based, it isn't necessary to know the microstructure of ground. It is just enough to know this road is right.

And then it is so that in the giving technology, well, which ... I have just told about, that this technology, which is called "Teaching of Grigori Grabovoi about Eternal Life", in the giving level you just may like through the way of connection, may not you?.. to the One source of eternal development, also through the way of straight contact, with the Physical Body of the One God, you can easy to have the norm only if you find the very right way of connection according to logical thinking, spiritual level, aspiration and work, and the action of Soul. So with the point of technology of Eternity everything has specific limits. An action of your Soul is absolutely concrete level of your personality. So spiritual striving exactly has a space where, for example, structure of action of Soul does not get. And logical level is expressed here as a union one, which is placed both in the level of action of Soul and on the striving level. So it is so that Love is the God's manifestation in this case. So Love is that the God loves everything, and He shows it, He shows with his action that the God's Love exists. In the giving case we see the God's Love. It is logically clear that when we see openly the God's Love, people often recover their sight, namely diseases ate treated: for example, eyesight and so on. And through the body - such a powerful optic, it like warms with such a tender God's Love, and the man is treated. Very quick mass treating may occur, as soon as a person, generally, in a big quantity, comes to knowledge of this information or just many people come to singular knowledge, but to this information, don't they? Why it often happens so that when a great amount of people, mote recoveries come from one singular right knowledge.

And when we see such a technology in action, we can make own mechanisms of governance, which primarily ate directed to such a way that our own bodies were such as they eternally develop, that means they must be healthy. And within this any phenomena, surely everyone connects it with the time, it happens so that a man does not spare enough time for himself, though he knows for a long time that some situation may occur. As a rule, many people know on that level of Soul, where Soul touches Eternity, what may happen with them. But not all people pay attention to this. And when you watch according to this situation on yourself, you see the way your own thought organizes.

Someone's screens of the thought are like projection, like a space of half sphere of Eternity, and when the information is perceived, the right decision is taken, isn't it?.. he passes the situation and influences it somehow. But someone does not react, and the systems ate more plane, he gives up them back, these signals or sees them somewhere in his dreams. And for your thinking becomes more diagnostic, that is though you know what may happen with the point of some negative, positive variants, you are desirable, your development structure is found so that the limit of thinking is formed in the way you put the sphere of Eternity, the sphere, on which your own event is projected.

You know, events are quite situation, a concrete luminescence in information, they exist in a concrete place. And even when I give a numerical series, this numerical series also exists in a concrete place in a form of luminescence. And when you in such away begin, for example, develop, suppose, there is a numerical series, and it norms, doesn't it?.. the governing numerical series, then this is, in general, just like an informational level of luminescence, like a corridor, which permits to chose, for example, the one right decision. I see it, tell about it and in this situation you do in the same way.

Conditions have changed; it is desirable for you to master best of all this giving numerical series, the way it works and then to spread knowledge with your actions. So the learned one, who can do it, he spreads knowledge, because now he can do it by personal constructive actions are, as he is in the structure of the right action and decision and at the same time can show everyone that he has made it even not within words. As an example, he made information, didn't he?.. the God saw it and gave it the next ones.

In the technologies of Eternity the next who gives is also the God. Surely, He always gives, but in technologies of Eternity it is noticeable very evident as made event is directed on infinite always comes to the God. It is so in any case ... For example, sometimes people think: well, he has made something and when it comes to the God, for example, what kind of decision will be? And here the decision always can be constructive, as a man is directed to Eternity, that means he does as the God does, and through the eternal limit of the Light the God meets it anyway, if people think that some decision comes, for example, slowly, though the God really see everything at once and right now, as I said according to the technologies of Eternity.

Just a union knowledge is fixed in the God's Soul in the giving case. And when you want to understand here what is Soul of the One God, who has the One Physical Body, His Soul is practically what in the giving case the whole World grasps as the World of action, which interacts with him on his free will. With the point of the God's Soul this is the own will of outer World. Why the God gives freedom to everyone? You see, it is impossible not to give freedom in the eternal development That means that a person should come to some opinion in a free way by himself, because this is the eternal combination of events, and the events are in an eternal amount. Consequently, if he performs some not free actions, than after some time a temptation appears for him, doesn't it?.. to do some, maybe, negative actions. So maybe for him they are positive, but for others are negative, aren't they? To see that limit, where they are positive for everyone is also the God's action. And when a person begins to move to the side of eternal development from this his way, he understands that the consciousness in this eternal way is an absolute level, where Love, that he has, also takes place, so he is the source of Love. And when you begin to see such a situation, you see that recovery, for example, of tissue, norm of events or there, we suppose, the governance of events, for negative level of events does not occur, this is the level, which primary was given to you by the Creator. So you see the primary point, which was giving to you at once by the Creator, as so in the capacity of independent work you will come to this system of comprehension. You see, this is the comprehension that takes place in technologies of Eternity. And when you see it, this luminescence will spread on all people at once according to my "Teaching about Eternal Life". You will pass your knowledge just from comprehension, here there is knowledge from comprehension. Well, you often noticed that knowledge is passed, somehow, not clearly. Try to identify absolutely concrete aims in the capacity of governance, and only, for example, on the material of today's seminar, urgent lection to perform the governance specifically concrete. So you will see how comprehension of knowledge about the level of eternal development gives realization immediately, well, for example, first of all in informational aspect, can almost at once give realization in a physical aspect and so on. I finish today's seminar on this point. Thank you for attention.

Translation from Russian into English was made by Permanent Creation.

DOCTRINE OF THE EARTH

The text of the work was first created by Grabovoi G.P. during the seminar the 4th of November 2003.

Once again, good morning. The theme of our today's seminar is my "Doctrine of the Earth." First of all, we should be considered in the given structure of the Teaching that the Earth is destined solely for people to live. That is, all the manifestations of the Earth in terms of, for example, the external structure - the structure interaction, right, with the outer space - in the structure of the divine development exactly are meant for the man to live. Therefore, considering the given structure of the Teaching, first of all, we can immediately consider the option, including saving the Earth from a possible global catastrophe.

For this it is necessary to consider that deep connection of the Earth, for example, with the human consciousness, with the thing that ate on the territory, yes, exactly on the surface of the Earth, say, happens in terms of, for example, a possible disaster, including the possible bacteriological destruction, because the Earth is very reactive to a structure of, for example, super-bacteriological type. That is the bacteriological weapons, in fact, can also cause the structure of the Earth's decomposition into fragmental constituents, that is, in general, currently has not been explored sufficiently.

So when we talk about the structure of the Earth, we must first consider the parameter of the man's development exactly in the direction of deep, deep understanding of the Earth in the Creator's structure. Because the Earth- it is also a part of the Creator, it is also a structure of God's development in terms of infinite life, therefore it should be endowed with the functions of perpetual development. And hence it is possible to consider the level of the Earth's eternity as a structure that develops together with the man. If we consider this position in the same way as the structure of the Earth including the direction of a person, it becomes clear that in my "Doctrine of the Earth," the transmission of information, training information, to be exact, is that, first of all, the Earth created for eternal life, then learning occurs almost instantly, even from the term Earth (?).

Here, it makes sense to consider the position of the World, that sometimes the term has a structure of, kind of, eternal, or infinite management. Here, in this case the Earth possesses exactly the term, for example, of eternal management even via the structure of the latter, for example, a vowel sound. Yes? .. If we say "I", we mean that, for example, "I" is a personality. If we want to, I mean, consider the structure - within your Consciousness, "I am -where I am," then the concept of "Earth" specifies the location. That is, the Earth is a place, yes, of human development. And at this such seemingly discretely divided word as "Earth," then we anyway always mean in the perception that the Earth and man - it is something united. That is, the man considers the Earth as the structure of his growth and development. And so in my "Doctrine about the Earth," we must first consider the spiritual relationship between man and the Soul in terms of perception of the Earth. That is, the man at the spiritual level perceives, generally speaking, the development of his Soul in the structure of the Earth. That is, first of all the Earth in the field of Collective Consciousness turns out to consist a person's Spirit advanced to cognizing the Soul. When we start, kind of, study the phraseology, it turns out that, for example, for the man to recover , let's say, restore his health, or to perform a resurrection, or, for example, to perform self-resurrection in the structure of interaction with the Earth, it is important to consider here the condition of the man's Soul of man in perceiving the Earth. That is, the man perceive the Earth as a spiritual state of his Soul.

Here's a very important term from the point of view of the fact that the Soul also has a certain spiritual state. And this level of spiritual states as well as the Physical Body of the One God, embodies that the man's Spirit, it in this case is regarded as a projection of the Creator's Soul, that is the One God, on the Earth. That's when you start, for example, copy out, well, or listen to these terms repeatedly, there are visible connections between the spiritual state of the Soul, between, generally speaking, a human Soul and the formation of the Earth. As soon as we come to this point of cognition, we get that the man's Soul knows quite exactly any parameter of the Earth. In principle, the Soul as it is, in general, has the knowledge about the World, but, I mean, that knowledge about the Earth is primarily located at the man's cellular level. man. That is, the man perceives the Earth as a very close, or we can say, the closest structure in terms of eternal development.

But if, for example, to extrapolate and to consider the position when there is some, for example, object that is developing in the aerospace, and it is focused on the eternal development, such an object is the Earth, then we perform certain elements the interaction with this object. That is just to pose the problem logically. And considering the way elements .interact from the God's point of view in particular, then the first element is the existence of the given object is in the infinite space conditions, i.e., under the circumstances where, for example, a comet may appear from any side, there may happen some unknown processes.

Then, it turns out that the object needs to interact, for example, with the man, yes, in order to find ways out of this situation. That is, the object, well, in fact, has the same development characteristics as the man.

And so when we, I mean, say that there are some nuclear tests, yes, some ecological problems happen on the Earth, then, on the other hand, we can consider that the Earth, for example, isn't it? .. allows the given structures - if we consider it as a structure of life on the Earth, the Earth is a living organism - to ensure that there are some elements of future development. That is, in this case the Earth behaves reasonably, allowing a person, for example, yes, to perform some action on the surface or there, for example, in the depth, right?... I mean somewhere beneath the surface of the planet. It turns out here that the structure of the Earth's development looks like a very reasonable structure related to the fact that if, for example, there is a very large planet, then some, well, structures of a smaller nature, there, let us assume for the Saturn, the interaction with the same comet, which is dangerous for the Earth, well, is completely unproblematic, in general.

It turns out that the Earth may be on the path, for example, of increasing its size, right? .. The easiest way is when people are resurrected, the sizes increase. That is, one of the statuses of the Earth's development is resurrection of, I mean, the gone. The second status if, for example, there some films showing that it is possible to work out some kind of weapon and the comet may be, let it be, yes?... stopped by an explosion. But in fact the Earth des not develop as a structure of a kind of explosive character. In fact, the Earth as God's structure gives only the creative development. And those development systems of destructive character, for example, macro nuclear weapon - they are more like a development structure of the human mind, mostly in terms of free -will. That is, a person can, of course, test some kind of systems, but for the Earth, for its real salvation the systems of global destruction are ineffective.

Therefore, I have talked more such a word, as a 'structure", as the structural systems of relations, for example, the Spirit of the man and the Earth, or, well, the Soul of the man and the Earth. In fact, there is a system of fundamental relations, which suggests that the Earth and the man are a unity in the plan of eternal development, at least, as a cradle, yes. Humanity and civilization, the Earth, of course, tend to situation when the Earth always keeps the primary source information, for example, in the eternal development.

But if we consider a simple model problem clearly, maybe mechanically: there is an element in the Eternity, well, let us suppose, there is some large sphere, where all the elements are eternal, then the center of the sphere must be eternal, too. And if we ask ourselves the question: where the initial level in the Eternity, which resulted in, right? ... the entire external structure of Eternity? That is, asking the question for the man, we come to the Earth. And when we ask a question, in general, in principle, on the spiritual basis, then we come to God, and at this to the Physical Body of the One God, which also interacts with the Earth, actually at the same spiritual status as the man; for God having come, right, for example, in the physical body, He also walks on the Earth.

And it turns out that in this case, it is the criterion that does not separate Him from the man at the spiritual level. That is, the Earth is the status of the spiritual, in general, partnership of God and the man. God and the man on the Earth have the equal development status. As it is logically clear that if we talk about the Moon, then people there walk in spacesuits, let us assume, and this is not natural environment for human beings. But if we want to develop in the direction of infinite space there, the word “spacesuit”, i.e. something that looks like a human body - at least, reflects its shape, right? .. and contains, for instance, oxygen for the beginning - that these systems, in general, still interact -with a human form. And it turns out that the task of all, so to speak, external attributes of development, let us suppose, of the man is also maintaining its from, in fact, the one, which is the closer to the human body, the closer it takes the form parameters of a man.

And when we study, for instance, the Earth from these positions, and at this we see that it still more spheroid, then we move on to the structure, more connected, for example, -with a man’s physical body. For example, we can consider that the head of a man if closer, is a kind of a spheroid structure, and to which the products of, right?.. thinking relate more - a thought is reproduced still closer to a spheroid form. That is, it turns out, at the material level, for God a material structure of thinking development is the Earth in the material manifestation. That is God practically thinks in such categories, as first of all, right?.. - the Earth is necessary for human life. That is, Gods thinking, in this case, becomes clearer, when we, for instance, to have peace, happiness, all over the Earth, and to have no wars. What we primarily mean is that God logically laid in the plan, that it is natural for a person who lives, to have some basis for life.

And further, while continuing to develop in the direction of the man - I told that the spacesuit looks more like a human body, than, for example, well, a spaceship in the form of a rocket - here it turns out that this similarity to the physical body of a man transforms into a structure of, kind of, spheroid processes. That is, we can see at a big distance that there are some big, for instance, optical systems in the form of star radiance, some planets more like a circle projection and so on.

And it turns out that the closer we, in general, in fact, come to that structure where God communicates -with the man, in general, giving him life, in essence, right? .. we come exactly to the structure of the spheroid interaction. For example, a projection of the distant future - it is a sphere for a man in the visual space. If we transform this projection into the structure of a particular object, that is, we transform these spheres into the structure of the landscape, the nature, we see that while overflowing into the object this area kind of descends, and we see, in general, the landscape. That is, we can see, knowing the structure of the Earth from the perspective of God, the man and the Earth itself how the future structure practically flow into the structures of solid systems: well, there, for example, in the mountains, in the grass, kind of, - and so on. But mostly, the easiest way to watch it is in the mountains, where there are hard rocks that are changing slowly. Then we can fully develop a very effective mechanism for restoring a human being and, generally speaking, creating conditions for his eternal development in the single, well, physical body.

As soon as we begin to understand how future processes that we see in terms of the development of the Earth, our physical body, and not to divide, for instance, the existing body shape with the shape of, let us assume, then, more spheroid - when we have already seen from the satellites, for example, in integrity, right? .. of the Earth. Here it turns out that the .flow is carried out by cognizing a human Spirit, anyway. That is the Spirit of a man who wants to cognize the future, sees the way the future enters the structure, for example, of the World. It turns out that in order for a person not to die at all, it is enough to show - well, just by thinking - a form of, let us say, the planets, the form of the Earth, so a form in general, in essence, of everything he encounters - but first of all exactly the Earth - hence, into his physical structure. A reflection - this is for God, for God's thinking, it is the interaction with the Earth. That is, God has created almost all conditions: that is, if a man puts one foot on the ground, yes, he has all chances, well, step there the eternal number of years. Why, then, the man possessing such criteria, still gets into the system, let us assume, well, of biological destruction? Here there is a dissonance exactly with the structure of thinking of God and the man. That is, if the man thinks by visualizing images which he sees as linear systems, then God does in a multidimensional system, naturally, thinks like ... Well, f1rst of all, like, yes? .. for the man there is, first of all, the Earth for his eternal development, exactly of the life, and only then he can go to a structure, a man, in general, approaches the level, when a man is walking on the Earth, and it gives him the means to development. This is a logical model.

On the spiritual level, God considers human development as the foundation of the Earth in general, as the foundation in the Earth structure. And so these combinations lead to the fact that one has only to get to the product of thought the same as God thinks, and we are, well, have opened here these so-called corridor ... system, i.e. the optical line is very quickly undone in front of you, and an inf1nite number of combinations, well, of normal development or kind of variable development are transformed back to normal. That is, the Earth - it is a criterion for the norm in terms of God, for the man. That is, if we consider this process as a process of normalization, we see that all the other species of animals, well, fish and so on, plankton, birds - everything that is on the Earth - it is a structure of varying the norm. That is, if somewhere there is a high radioactive level, there birds, also can get in the phase of destruction.

And it turns out that if we consider the lines of development, such as the Earth and vegetation, as a system of normalization in fact, isn't it? .. as the direction for human development, then, in general, it turns out that way, in general, of the man, first of all, is directed to seeing the structure of your, for example, eternal development, even when you look at some ordinary object of the earthly nature: a butterfly is flying there, or, let us suppose, you look at the grass. That is, it is the coordinate of the fact that, if here, for example, there is grass, right, it means that life is possible here, at this point. And with the development of, for example, into other Worlds, when normalization of the structure of health exactly - that is, a person must be healthy, but, for example, to develop into any other system then the criterion is again very simple combinations, in general, of the earthly life, that is why there are many experiments conducted now on creating the conditions for the life of plants, for example, yes, in a Flight. Exactly the way that I am talking about, they considered the development position, but my point of view, it is precisely because it, I mean, the structure of animals' development in freedom. And in this freedom away from people and from God, it, kind of, led to the fact that the intelligence of these animals was such that, for example, cows stopped feeling afraid of, yes? .. well, the man and began to develop towards, in general, the normal interaction in terms of and it turns out that this is the very item of, kind of, the external structure overflowing, to, for instance, plants, well, animals. And approaching, for example, a person - that is, the outer landscape of the future evolves towards you - and when it comes up to a man, the development stops here, because there is already a man who 1s originally created by God to have the eternal development. Here the next level of interaction appears - here a person begins to affect the Earth by his structure, that is, by his personality. And, in general, in the logics it is transformed into the fact, that the man can really affect the environment, that the system of global destruction is really possible, for example, isn't it? .. and so on. And the less this possibility, the more the man understands the very deep internal structure of the interaction between the Earth and the man, that is the Earth and God, and the way God, in general, I mean, showed the Earth, so to speak, right? .. that the Earth is supposed to do while interacting with the man. Then it will be easier. That is, there is a general level - God has indicated the development, for example. And the next level - that this development should be covered by both the Earth and the man.

Then it turns out that if you compare it, for example, well, with some navigation map, so, in the development, in the means movement, for example, of the infinite development, it turns out that it is enough to get a command simultaneously, that is to get it right - the Earth must get it right, yes, and the man - then we get synchronization in development. And this structure of understanding, in fact, first of all, it results in the structure of animals' development. That is, in the animals' understanding of almost the same what the man understands and knows - well, maybe not fully, well, not necessarily technological aspects like machines - but exactly in the understanding of the same. Why, for example, well, in ancient India they first of all considered the animals that are close to man, yes, let us assume, the cows. They were given opportunities, well, in general, of co-development, in order to, in particular ... Maybe not uniform criteria: the, let us say, criteria of the eternal development, criteria of, well, reasonable development- and so on.

And it turns out that the very element of interaction, it is for God is a kind of canon, when the man, then, comes to the fact that he does not destroy, yes, the natural habitat, well, let us assume, of the animals, he creates the conditions for, well, the life there, for example, plants, forests and so on. And here we obtain a sufficiently balanced level of interaction in terms of diagnostic, well, systems, when a person realizes this and does it as if on the physical level, he, it appears, within the frames, in general, of the same team even from the logical point of view. For the Earth it is important to grow plants, for example, yes, and for the man it is also important. That is, the team 1s identified as, for example, a particular plant photosynthesis. This is the team, for example. Yes? ... If you look further, what is going on there in the species of animals? The animal that develops in the given natural habitat, what for does it develop? So, why do we need wolves, let us say, yes, deer and elks? If we consider this level of interaction, then by assigning the task to do so that the wolf would not be aggressive to the elk, yes, let us suppose, that it would not eat, then it turns out that the man should provide the wolf, let us say, with some, perhaps, synthetic food. But, it means that when we consider such a regulatory function, well, of the man, then we come to the idea that, in principle, the man does have' to interfere in the management structure, I mean, yes, of the external nature at a certain level of his development, in order to, well, there would be correspondence to that fact that all systems, even, in general, of the aggressive level, develop synchronously and towards goodness of development, kind of. That is, the wolf does not eat the deer, the elk well, because the man fed it, that is, the man provided some conditions. The man is given, I mean, to bring some of the aggressive systems, there, yes, to the system, in general, of the parallel development in the form of non-aggressive level. Then the person begins to learn how to exist in the World where there is a potential global destruction, for example, Yes? .. How can he transform the technology interaction, well, with the wolf, when the wolf does not eat another animal out there, into not exploding a nuclear bomb.

And here, by the way, there are very specific techniques, that are designed not only to ensure that a bomb would not explode, right?

.. but also aim at healing well, let us assume, the fourth stage of cancer, healing incurable diseases.

For example, in some papers I conducted management, where, for example, considering such interaction as the structure of the fauna - one of the systems out there, yes - a predator, the other, let us assume - the one that is not a predator - it is very easy to see how, then, the influence happens, in order to prevent cell mutations and not to let the malignant process occur. That is, the analysis of these processes gives a very strong recovery system. For example, to recover, yes, the man with the same disease as, well, the fourth stage of cancer, it is enough to mentally bring himself there in the structure: there is a wolf or some predator, it does not matter who is there, there is some predatory dinosaur, and the other animal, not a predator - and to try your best in order for the predator not to eat that animal, which is not a predator. That is, with an effort of will, the management system to try to find yow:: place, right, to prevent this, in general, aggressive manifestation of the system. Then the recovery of sufficiently serious diseases happens.

That is, one of the elements of the management, for example, often described there in some techniques, that, for example, a man comes to the tree, if it finds the correct location between the trees, right? .. then he is cured. Here, in general, the same system, but at a weaker, perhaps a signaling level: if one tree interacts harmoniously with the other, the man, yes, enters the recovery phase. This, in general, is a technology that is presented simply, in general, at the usual level of plants, animals and so on. And it turns out that for humans and the Earth there is again the same, but now an intellectual, problem. That is, to create the development norm system for any kinds of, well, biological systems and thus to create conditions so that, for example, the gone, they would obligatory arise. That is, to create the following development system based on, for example, the existing geometric systems of, kind of, deterrence.

That is, logically where, apparently, it is possible to understand that the Earth may be increased? But recently, in general, over the spent time - about two and a half years - the mass of the Earth, well, it, kind of, even ... The concentration on the optical range has intensified, that, first, there is much external matter on the Earth. But here, even if you calculate the full range of increase in mass of .the Earth by satellite monitoring, it is incomparably greater than, kind of, all the external meteorite showers, and so on, at this, the width of the Earth increases literally, well, for millimeters per year in some areas.

And it turns out that if this process is now considered as a whole, it is clear that as soon as a person tl1e.re or ten people there are resurrected, the volume is still increasing. That weight is connected with a real spiritual act of a person and simultaneously with an action of God, on, for instance, restoring life. And when we consider this process as, let us say, that there are the bones of ancient animals, they exist internally, i.e. after a certain time these bones are not found. For example, there, minus five thousand years in some regions, for example, there are practically no bones found, let us suppose, of people there, or some animal species, when it is kind of known from the historical position that there were, yes, for example, there at least, I mean, those animals that ate registered there in the nearest habitats. And there's a certain angle, that within a certain time, the Earth itself restores the structure of these animals, well, for example, there in the next space, perhaps, in another place. In terms of development, for example, of different types and systems, such a mechanism is quite well known in terms of the existence of the so called tissue development occur, but in a living form already. That is, tl1e Earth has a recovery system of the man or other animal species in a living ductile underground rivers. That is, there happens the transfer of large mass of material, well, and there, in general, the elements of, say, bone form.

And when we, for example, begin to explore this mechanism, we see that the system of colonization of different continents here and there, and so on, of the objects, has, in principle, a different perspective of development, rather than a mechanical one: not only where the movement of the people themselves or there some kinds of animals to different regions, but, actually the mechanism of the Earth's development with respect, for example, to the future process of the eternal development. And here it turns out that once we begin to understand that the Earth is quite an active level of interaction, so, for example, with the man, with the future development of the Earth itself, then we understand that some earthquakes can be, for example, profiled, if we know why, kind of, we have, for example, some kind of a shake-up, well, for the Earth at the given element of the temporary planet development. Here it is sufficient to leave the planet development for some time. And it turns out that after a certain time, there, for example, where there are no bones, so here people are restored. That is, they are, it turns out, just arose over some time.

And when we, for example, begin to analyze this process, it turns out that, for example, for the bone structure of even a man - if we consider structuring, kind of, the bone tissue during the a man's growth, well, during the development from the fetal development, in terms of even initial periods , well, the first few weeks - then the bone tissue grows according to the same principle as the Earth evolves to infinity. That is, it develops so structurally at the, kind of, molecular level. If we compare these models, then, let us suppose, even after taking the layer wise spectrogram, we get the same thing as the optical range of the external space. At the moment, if we look inward, i.e. from the outside, at the tissue, then we see the structure of a human being. That is, it turns out, that the man on a mechanical microelement level always comes on his own form. But in principle it applies generally to any other species, I mean, that exist on the Earth. As for other, let us assume, systems or other planets if to speak certainly, we can only say, for example, after sending a spaceship or developing your own spiritual system of diagnostics. But, in fact, at the spiritual level, any system of the World, develops exactly like this.

It turns out that the system of the Earth' development in terms of, well, maybe, some new knowledge of technical and scientific nature, it allows you to interact in terms that the Earth is also, in general, the structure having a system of perpetual development. In contrast, for example, from the man, if the man does it to a greater extent with his Soul, yes, for example, raises another man, the Earth does it exactly by the level, which is physically manifest. For the Earth is a form of thought of the One God. We can see that the Earth can grow specifically, for example, in a physical body of a man. And when, let us say, one of the elements of the human restoration, there, in quite ancient civilizations was the element of the body parts interaction, let us assume, there with the ground: wound, let us suppose, on the left leg could be cured if to cover with the earth the right hand and the right leg. And moreover, then, for example: sometimes more than the right hand or the whole right leg. That is, there were some very serious techniques of interacting with the Earth. Many different diseases are cured by such different elements, well, now it remains, that there are some mud baths and so on. That is, there is something that acts, kind of, only on a number of diseases, and only in the complex application. The interaction between, let us say, the Earth and the man in terms of this man's life is the main aspect of the specialization of the Earth.

If we consider, let us say, such an element, that here, let us suppose, the sound of a woodpecker, that knocks on wood. Then it turns out that the sound annoys a person more than, for example, the some other, well, sound, because it is clear that the woodpecker, in general, breaks the tree itself. And when comparing, for example, the sound with any other, it appeared that the sound has a very specific and direct effect on a human gene straight, it kind of passes over even the structure of the physical level of a person, because the person initially knows there is some process of destruction going on. That is, any process of destruction, which are rendered by hearing, seeing, they have a direct impact on human genetics.

When we consider the process of interaction at the spiritual level, then, there are enlightened people - those who possess the system of clairvoyance - they can as well perceive everything happening far away from the outside, right? ... at the level that is reflected there in their Consciousness.

And there is the following problem: for this, like, yes, not occur, first of all. That is, the development of clairvoyance, so, for example, via a genetic system, maybe, occurs, in general, through the auditory parameter. So, we have to in such a way, as not to destroy the tree, we must know where the woodpecker is, yes, and, for example, to frighten it away. Hence, it is clear for the hearing. But if the woodpecker is very far, how to come to an agreement with him not to hollow the tree, if there is his food, let us suppose, right? .. There an element, well, of interacting with the target appears.

Let us suppose, that there is some state, which has an aggressive nuclear weapon system, and, in principle, a system of, well, attack, if not nuclear weapons - a system of state development, permission, I mean, to restrain the political system within the necessary track, there is population and so on. This is, in general, a task similar to that task, that if not to drive away the woodpecker by the hand, then the political system can also be influenced by, kind of, management through the managing clairvoyance. And when we start to compare the nature of some elements of the system, for example, well, politics, social development, then we see that the Spirit of God Himself, the One God at this, is manifested in very small elements of the Earth's reality in term of the situation when each element interacts with the entire social system. That is, any micro particle of the Earth interacts with the social system of, well, the processes of the future of the given land, of the given territory.

And when we begin to enter it this mind of the Earth, we see that the Earth affect, among the rest the level of how relations, there, for example, of the use, right? .. of rent are built. In some areas, they are very simple with the same legislation. The so-called economic paradox: that is especially in the beginning of the XX century, where, well, they are seemingly identical legislations, identical systems of rent relations, yes, in capital countries - some territory still cannot for a number of various reasons to be developed, others are very effectively developed. Here we have precisely the influence of the Earth1s structure on the parameter, for example, yes, the of economic process of regulation by, as it seems, the man. That is, the Earth intrudes, among the rest in a reasonable manner, well, in general, from the point of view of the man and from the point of view of animals that live there, into the system of more advanced, it would seem, well, systems, well, economic systems of human development. So, it appears that there are parameters of management given for the Earth, connected with the fact that a more complicated system can also be managed. And, on this basis, a person can consider quite unequivocally that if there is a more complicated system, well, more, kind of, for example, developed, so to speak, then with them, first of all, he can always come to an agreement with the more developed ones, because the more developed still imply a more peaceful, kind of, a more humane attitude towards the system, right? .. of development of the man himself. Because first of all, they see the man who, for example, then, well, let us say, does not have, yes, such a technical system of development, although, from the perspective of God, the man can manage, generally speaking, any system in the same way like any system can manage any other one.

That is, this exact level is very process, in general, in fact, of the growth of human Consciousness into the Consciousness of the Earth. Right here in my 11Doctrine the Earth,11 there is, such a, in general, special level of human Consciousness development as the development of Consciousness within Consciousness, where one level is inseparable from the other. And if human consciousness begins, therefore, to develop within the structure of the eternal development exactly - that is, by Consciousness the man realizes that it is necessary to be done, then the man should have perpetual development- then it turns out, that the Consciousness of the earth begins to develop intellectually in the way of interaction with the man towards the perpetual development. Then, very many processes are resolved by simple interaction of thinking and the man, well, suppose there, into the Biblical, yes, episodes when they sea moves apart and so on. That is, the interaction of the Earth and the man in terms of eternity structure as it is - it is quite available, in fact, technology that can allow regulating the climate, can allow regulating, for example, well, the irrigation systems.

And so, in principle, in fact, for the perpetual development of the Earth it is necessary to understand exactly the organic structure of the interaction between the Earth and the man. For example, if we take some of the ancient civilization, which, in general, developed and mastered, for example, desert areas, which became deserts now, or they initially began to develop, but then everything turned into a desert anyway. That means that the primary element of the development was, in general, correct, since they were able to irrigate, but further, kind of, consumer, right, attitude led, for example, to the land devastation. Therefore, from the perspective of the Earth the man, of course, must give as much as possible compared to what, for example, he can, well, take. That is, the principle of human development on the Earth is to be able to give much more to, as it actually happens, to yourself. This is a medium on which there is the man himself. And, if at this moment he give more, he lays the future management structure, yes, for self-development Technologically, this complex of technological, kind of, clearly expressed by the example of interaction between the Earth methods in terms of the given structure of my teaching, Doctrine of and the man, where, as it seems, inert level in mechanistic understanding such as the Earth, exhibits very strong intelligent functions of, right, management.

And here we can thoroughly consider such a process as a the Earth, here we can consider a number of very specific techniques aimed at the human ability to interact in the direction of the future perpetual development with the Earth at very precise criteria levels, it means, knowledge. That is, he knows that it is this way, for example, and he gets the perpetual development. That is the first level: knowing that the structure, such as God, right? .. of the Earth and the man is manifested for example in the interaction with future events through, for example, plants, we may say that it is enough just to look at a certain angle at some growing plant - we, via the conversion system, get to the future events, but they seem, kind of, to be very soft and favorable to us. That is, if you look at the future, it is easier to originally look through the plant structure, so, for example - that is, the first aspect, which the given practice of cognizing the Earth gives us, from the point of v1ew, for example, of human Consciousness. But if we look from the perspective of Consciousness of the One God, here it turns out that in order to adapt your consciousness to such viewing from God, we should see his thought, i.e. thinking. The idea of the Earth for him is the same thing.

Then it turns out that ... And in thinking it is possible to develop any constructions, that is, any in terms of eternal development as it is. And it turns out that just the way the man thinks - he may think, in general, completely positively for his future or, there, at least, to determine the future positive criteria - then, the same way God can create the most positive future, because for him it is also thinking, that is, the infinite level of development. And the variability of possible development means that the Earth is a state of mind of God. When you start thinking this way, you can see that a thought depends on a spiritual state as well, that is the state of the Spirit of God. It turns out that God gains human qualities, associated with the Earth, i.e., exactly the ones coming from the eternal level 3a the Earth. Then God becomes closer. That is, understanding of the Earth, and hence contributing to its perpetual development- it is also, in general, drawing God closer to your vital problems. Logically, of course, it is like this: if we know the laws of development, there, let us take soil, good? .. we can plant more wheat and so on. And here is it happens that ... Well, even more, when we know the way the soil develops from the point of view of God, then we can make the wheat, for example, yes, infinitely growing. Well, some kind of system that allows the development of, for example, there is an infinite number of products from the same place. 39 That is, here it turns out, that we can use quite well, for example, the technique of developing our Consciousness toward cognizing the Earth, in order to almost perfectly implement these technologies. So, unlike, for example, may be, from technologies aimed at on some external civilizations, planets, and so on, it is exactly the Earth, where you can conduct manage, which will be traceable by you as your practical work. That is, if there is a wood drying out somewhere, what is necessary to be done for it not to dry up. If there is a pond located somewhere outside the ecologically normal environmental situation, then, how to clean it by, let us assume, your Consciousness.

That is, when you start working on this aspect, you can see that the Earth already gives you more, and much more. So it seems, well, there, you have invested, I mean, for example, very much, or more than, well, in another system, and further on you get much more anyway. That is, the returns from the Earth to you in terms of, in general, a certain promotion, a certain softness of events, as that plant, which is, kind of, a gateway, a filter for the softness of the future. Similarly, the Earth - is the filter, it is that certain, let us say, prism, which provides the future.

And when we, therefore, consider this structure exactly as a support system, we understand that it is clear, that in this system there obviously should be no problems. That is, for example, the system of providing the man with food and drink it, in fact, in the first place, for example, there at some kind of spaceship must be, in general, in good order. It is natural that the man should invest into the system's being completely manageable, that is, it must have infinite resources, it must have a structure of infinite development. If you go by analogy, for example of this process, it turns out that when you, for example, consider the process of the Earth's restoring a system, that is, the Earth has its own recovery functions, and if the process is speeded up, then you may get quite fast regeneration of the Earth itself, soil, plants and so on. But, in essence, from the Creator, from the One God, there is the process of restoring any destroyed information object, or the converted one. That is, if there was a plant and it was destroyed in some way, perhaps even by a change of seasons, it turns out that - according to the Law of the Creator - it should be restored.

It turns out that if we look this way at the time axis, then the element of matter transformation - is a structure, in fact, of time management, but from side of the Earth. If we want to restore any leaf out there that has ever existed in civilization, we need to make this leaf, well, kind of, transfer it to the center of the World, to give it complete freedom of development, and then around the leaf there will appear the space of, first of all, the Earth, and which will be completely free, and then the man could, for example, occupy it. Right?.. So, in principle, the development of civilization, it happened almost this way: there were vacant lands, the man occupied them. That is ... But, if to be precise, this is what was the primary norm of the World. That is, the exact problem was set as if everything was indestructible. Then it turns out that the man develops the new land, yes, the new territory. As soon as the territories are filled, the person should be able to reproduce if by means of his thinking, or ... They certainly increase, those territories through the technologies of resurrection, but they increase exactly for this person, and the further development implies the presence of the opportunity to expand the range of his development in the person. It is even more likely that if, for example, the natural habitat may be extended to a passive management system, in this case we talk about some spiritual habitat of development, when the man sets problems for himself and develop them himself.

Well, let suppose there is a simple problem: how, then, to combine the space between the Earth and the Moon, right? .. For one thing - you can, of course, fly there, but for another thing - if you collect technologies and develop the Earth and oxygen around you through consciousness, then, in principle, in order not to disrupt, for example, the peace of the same Moon and not to let it shift, it is necessary to understand the gravity parameters, and understand that any other external body possesses the functions of inviolability given by God originally. Therefore, all this external structure of the vacuum and the system of, kind of, location of distant objects- is a system of inviolability of external systems or internal systems, isn't it? .. -with respect to the macro object. That is, in the Universe, for example, the processes when, let us say, some macro object, let us suppose, it is somehow, well, gets destroyed, are extremely rare.

Consequently, some of these macro systems, that is, the outer World, that is external, kind of, optical space, it is constructed in such a way that we first solve the problem of non-tangency, that is, equality of development. If we solve this issue through a logical plan, right? ... that is, we say, "Yes, we can build up a large amount of the Earth, but we will not touch the outer planets. We will create a gravitational parameter by our Consciousness, which will hold the data of the planet in their place". Then we can quite quickly, for example, develop the Earth to the desired number of kilometers or, well, tons. Yes? ... That is, this development process is mostly limited exactly by the structure of unrest raining, kind of, elements of the free development of other information objects expressed materially. Because if you pay attention to, let us suppose, the fact that you think about some matter, and the intersection of another idea -with this idea, than there a kind of resonance occurs, yes, and this is often impossible. That is, you can do so by a will effort. There, for example, you have to go to the store and at the same time you are thinking about something else, you can create another idea at the free level by the substructure of your thinking. The same way with God: thinking, He, for example, created the Earth, but the intersection of ideas here is impossible, you can create only a substructure. That is, to develop matter in the structure when another substance has the same semantic status. That is, based on the example of our thinking we can very easily consider how physical matter of the materially expressed World develops. And when you start here, by analogy with God to reason, it turns out that there is a concept of kind of external vacuum, yes -it is very dense, kind of restraining the balance matter in terms of this exact process. And it turns out that to get the structure of the Earth there in the vacuum, you only need to include the balance parameter by the Consciousness. And that's all. Well, it is more like physics, of course, these exact arguments, but nonetheless it is exactly the way per se. That is, it is possible to make specific management system.

Let us assume that in order to grow an organ, right? ...within a person, you need only to create a structure so that your action will not affect, well, the interests or freedoms of other systems, external systems. And for that you only should know the structure of the organ and to try to create this organ just the way it is originally from God. And when you, for example, create the organ exactly the way it is originally from God, you get really, in general, stable management system, which is similar to the way, for example, isn't it? .. the Earth develops, according to the laws of the Earth's development you may synchronize quite well, for example, I mean, the development, for example, of some organ or, well a man's resurrection “'With the fact that the Earth is, in principle, develops according to the same laws.

That is, reflecting the structure of kind of external management, which comes from the Earth, from the structure of life, that is drawing to the man those attribute the Earth's growth - it means plants, it means, well, the normal natural world - thus you contribute, in general, to the man's norm, and at the same time, it appears, to the norm of the Earth. Because the man's norm simultaneously affects the norm of the Earth. This is a very significant process, and, in general, people determined quite long ago that the interaction with nature as possible - at least at the level of thinking out there, even, by the way, by watching some report- has a beneficial effect upon the man, not to mention the fact that it really helps nature and develops in this respect

Therefore, from the point of my “Doctrine of the Earth,” there are a lot of technologies of interaction with the Earth itself, and, in principle, the interaction itself. The deeper and more fundamentally you consider is the presence of the Single God-the Creator here, the more precise system of management you will find exactly according to your tasks. And as far as it is possible here, consider the following aspects. So, considering those arguments, those arguments and those examples, specific tasks, specific management practice that I have shown you now, I mean, concerning your task, try to do the following: just as a kind of concentration, just to keep this task in mind. And you will see that even listening to this, for example, lecture, by means of simple, seemingly passive containment of the task in memory, leads to rapid resolution of this task as well. That is , generally speaking, this tool can be applied to other lecture systems, but exactly in the structure of the Earth, you will see how the Earth and all the statements about it just helps you, very serious and often radically. At this I finish today's seminar. Thank you very much for your attention.

Structurization of Consciousness Method of construction a dynamic element of reality in perception

The text of the work was first created by Grabovoi G.P. during the seminar the 22th of June 2001.

Good day everybody! Now I begin the first part, based on lectures. The giving lesson will be formed in the next way: at first the lecture, about an hour, then the answers for questions will follow. The next part of questions for the next lecture for example is better to keep now.

In the part of the giving lecture about my System of Salvation and Structure of Consciousness, which deals with prevention of feasible global cataclysm, I give a method of Structurization of Consciousness based on the point that you form an element of reality as such optical display on perception that you make dynamic the giving element of reality, The point is that my methods of Structurization of Consciousness - this technology may be: either step by step that is when you learn you get a new method every time; or each of methods may work by itself, especially in those cases which deal with salvation from a feasible global cataclysm; and so it is necessary to note that each of the methods is also autonomous.

So, the giving method is based on the technology of Perception. It means that Consciousness is considered as a system of governance, when the perception is considered as an element of Consciousness, but the aim of the giving method is to govern that is specifically to rescue from a feasible global cataclysm by giving dynamic properties for perception in the optical range, So, the principle of this method consists of that at first you consider some element of the reality as the element perceived initially. That is at first you should realize what your perception means for you: if you possess some information about feasible global cataclysm or, for example, do not possess. The case is that exactly this first part, the prevention of just feasible global cataclysm, actually this is sufficiently a narrow specialization on governance, and that is why this part should be divided as far as possible in future, though it is also aimed at the things you can do through it, well generally, any governance, but perceive as a narrow specialization.

And so, in distinction from like a background level of governance, in this case the initial level can either be displayed in perception, that is you may get some signal: that there is some problem, at that from which global catastrophic consequences are possible - either nowadays or in future; or you may just set a task exactly the searching one and in the searching task to work over that in system so that to prevent the feasible global cataclysm. Therefore here the situation consists of making the signal of perception dynamic in two positions, for example: either there is a signal, which exists basing on a certain outer problem, or you deal with a system level of governance, and then the principle will differ here, so, the first principle: when you respond to the element of outer reality, it means the certain information occurs in your perception, exactly your information; so I mean this information is not from TV set, for example, but exactly the information in your perception, or from some means of physical information; so, you get the information about feasible problem, and in this case the dynamic of governance keeps within definite axiomatic principles, describing which you ought to know that first of all it is necessary to distinguish that segment of information which means this in optical range in such a way that you could immediately perform title governing performance for failure of this element of reality, so to prevent the feasible global cataclysm, Consequently, basing on this, the further logic of development is so, that first of all you realize an impulse from your Consciousness, which carries exactly the main point of your work; so the task is not to fix, naturally, the feasible global cataclysm, but really be able to prevent it. Here the conception "really" consists of that you have to do such governance at once, so that it -will be stable in future time interval, that is it does not happen in future.

Consequently, this is the key position of the giving method in the case if it is an outer element of reality when you make a future stable on positions that the feasible global cataclysm, for example, does not occur there.

Then the situation is already simplified here now, so you need to make future construction dynamic. So that the future construction is dynamic in your perception, you just prolong your perception in the future field, but in such a way, so that it became stable, namely fixed, Prolongation in future through fixation in the past - this is such an axiomatic level This level is defined with that the first axiom, past, for giving the dynamic to image of perception, consists of the elements of beginning future, that means you distinguish some element for perception where the future begins. You can do it in elementary way in perception because you may distinguish the field of beginning future as a quite real field.

The second axiom consists of that the field of beginning future is, of course, your own governing system. But the axiom points that your governing system becomes a reality when it developments within the aim of the Creator. The aim of the Creator is creative harmonious development and you construct the reality in this way.

And the third axiom points that the construction of reality takes place in canons of eternal construction in that case if you distinguish the eternity in your own consciousness. It means you have to find the element of Consciousness, but consider Consciousness, for example, when you look at yourself, as your point of perception is in the physical body. As an example, the Soul is in the physical body, and you look at yourself from without, but this is as you look at your appearance exactly from without, and right now you will find the eternity in your Consciousness - this is your o -n task. Such is the task that as soon as you make up your mind to find the technology of eternity in Consciousness, you are sure to find it on the first impulse of perception, and this is impulse is a dynamic impulse.

The case is that the giving lecture will specially be complicated in the way of apprehending, and so it is better to listen to it for several times on the Dictaphone, for example, and follow the logic which I state now. And when you discover the eternity in outer perception - this is a dynamic characteristic of future governance. So it means that dynamic is just what you find the eternity in perception, by outer vision of yourself information. And as the presence here is compulsory, because you are the manifestation of eternity, it turns out any case that you save from the feasible global cataclysm.

If you convert the phraseology I present to structure, DNA by vibration structure for example, we receive exactly DNA system plus element of future development, which is based in a human tissue generally. That means that micro molecular level or DNA level is an information of development, which exists, as an example, in a form of microscopic researches, and manifestation of future is a human tissue in molecular manifestation, namely in a form.

Consequently, the future manifests as a man, and according to this - on information I mean - and according to this it is quite easy to find a system of governance from this point, if you know the way a man moves. It is laid in a definite automatism on the level, will say, where it is not necessary to use the control of Consciousness, where the elements of collective Consciousness are assembled exactly in that aspect, that the element of governance, connected with eternal life, bases on a human form.

Then you work with this technology by yourself, you will see that the position of eternal life is manifested in such a way that the eternal life is a compulsory condition for infinite functions of future processes. In this connection when I talk about such conceptions as "eternal life" or why the Creator gave the question in this way firstly, we receive these elements are such that actually they are just laws of information distribution. Let's invite those who have just come here. In view of the fact that we consider elements of eternal life as elements of development in this case, an element of eternity and its image is a man itself. So when I said you look at yourself and see the element of eternity, this is because you are manifestation, but in the technology of development of Consciousness this manages in the logic of Consciousness development laws.

So when you govern through such compositions as eternity and formation of future events, here you should consider as an independent work that these are phenomena of the same kind and phenomena, which follow one by one and intercross on levels of perception.

Consequently, the technology of creation of governing impulse on the basis of giving the dynamic for impulses of outer nature consists of building by yourself this technology consistently or just adding impulse of governance by perception yourself both in presence and in future.

So the technology is quite simple: to perceive yourself both in presence and future at the same time; and this impulse will govern the reality firmly in such a way that the global cataclysm will not happen. According to the system of governance from static image, that means you perform in such a way that you regularly look through whole reality and search those elements of governance, which create stable reality, preventing, for example, feasible global cataclysm in future or you deal with information of the past, so that the cataclysm does not happen in future. Systematization of this searching must consist of stable direction of your search for real preventing feasible global cataclysm; herewith you did not fix possible negative elements neither from the past no from the futur.

So information of negative aspect, which dealt with some feasible internuclear conflicts, still must be governed from urgent time of governing. It is well known that there were situations when internuclear conflicts could happen. Consequently, this information must not remain as information fixed for the past.

Anyway when you govern with the point of your searching systems, will say, when you consider reality and govern from yourself, and give the elements of perception dynamic characteristic, then you have to govern certainly the whole amount of information.

As an example, if you make a governance even in remote future, suppose we take one million years forward, all known information about possible conflicts, which could happen, is contained nevertheless in this part as Caribbean crisis for example. And in this connection this information exists in common amount of information as an element, which reacts on your governance in future. Logic is quite simple. You should be more attentive now as the logic I give on the level of logical governance, crosses with the level of Spiritual governance.

Here the Spirit may occur in different ways, capaciously, as this is your personal task, task of everyone - to prevent feasible global cataclysm and to spread this knowledge. And according to this the technologies of cognition, which will happen during the process of studying this part, they can be absolutely universal; but the main point is that they must be directed at only one aim - exactly prevention feasible global cataclysm through your own searching system, that is when you put a task specifically and search for what can be a reason; and how today it is necessary to work, so much as: an hour, second, minute or half-day; or maybe to work some hours constantly so that the cataclysm does not happen. Then you certainly have to govern the whole information of the past right here.

In this connection a series of axioms exists. They are quite simple in logical perception, but the whole point is that the education I give is Spiritual; the Spirit must be so that to prevent feasible and global cataclysm in any case, and not only global but better to remove any problem to the side of optimal decision. In this connection axiomatic level, which follows logically from optical laws of interaction, must be researched and understood by you: where perception turns into spiritual basis, why governance exists, when there is a real result, but there is no an intermediate oral level, for example. In this connection the optic of governance, is what now I am talking about optical range, which is understandable by everyone, must join on the level of cognition with the range of development of Spirit in compulsory way. What the development of Spirit means in this case? And in this case, as a rule, these are the elements of consciousness of the whole reality.

I put the elements of compulsory consciousness in my systems of concentration on every day. In connection with that the giving methodology is directed at that you govern independently and have your own instrument of governance, I just enter an independent level of development as the level which is necessary now; so this is not usual logic of thinking, for example, or you, suppose, consider the process in such a way that you have a certain informative constant from a certain source of information, which characterizes that there is a development, for example, and it is either off-speed or very fast. So, the main point is that it is you who busy yourself with any... with any element of reality, including development of your own Spirit and your own Soul according to purpose governance.

And so when I talk about the governance on optical range, as an example, join of two optical elements... suppose, I can examine a video camera or microphone as two optical elements of perception, just to convert to some signals, for example, I close my eyes and examine that they are really situated side by side; but according to interaction of these two signals I can get the Spiritual development in interpretation: what kind of objects they are.

The case is that the task of everyone is to research a field which does not have known characteristics - this is the main thing that is needed for salvation in any case; and also the understanding if any field which has known characteristics. In this connection it is well I known what, for example, the video camera and microphone are. I And the meaning of Spiritual development in the aspect of governance keeps within that task that I consider the elements of Spirit as some displayed creative elements with cognition of microcircuit of this card or mathematics or physics or some image.

But when you deal with unknown systems, in this case two optical factors, i.e. two optical instruments, which are side by side in your perception, conditionally as two lamps, they need governance, including the essence of their own cognition.

In this connection the Spiritual development is passing ahead future, which forms its own shape element by you are in these two elements of reality at the same time. Well, in ancient Yoga, for example, elements of development, which were connected only with a personal, suppose, development, consisted of those that a man had to realize himself in objects of external perception or his perception in one of stages. Analyze this phrase as an independent work here: why I said external perception or one's own perception? And when realization happened, the work was considered finished, because influence on reality occurred: suppose, they could thaw snow, i.e.raise temperature and to govern the reality absolutely evidently on a physical level, or, suppose, to revive themselves. Some news-reel shootings exist, which date back to the beginning of the century, when a yogi, suppose, gathered himself simply after he was burnt. There are totally evident events which are made and even recorded.

But the task of my governance, for example, is not to reach a new level to fix a negative status. It means the sense of salvation is to follow in any case such a logic of development, so that at first a human body must not be destroyed, but if, for example, it has some problem, it is necessary to recover it. So the principle I state now, concerning the researching a dynamic structure, which is a dynamic structure of the future, consists that there is no need, suppose, to traumatize something specially, or, suppose, cut a finger and to recover it in a moment.

The case is that the sense of governance is to do so that those people, who do not know about such methods, for example, they were up in spite of this on the level of transmission of information with your. So, the point is that the governance ought to be for everyone. Another case is when everyone can do it, but and then it is better to follow the ideological principle: first of all, not to injure harm, then to recover.

The clearing I have entered has a principle meaning so that to see dynamical structure of governance of future characteristics, and this dynamic structure is that you constantly reproduce your own recovered image and recovered according to the parameters from the past too. That means you have to accumulate your past in such a way, so that, for example, the fact that now you are, for example, in current time, is a result of those that you recovered your information in the past up to the level you need. Consequently, if you recover up to the level you need, the Caribbean crisis does not happen.

So the logic is very simple here. I brought this part of the lecture to this logic through the level, which has construction of various kinds from the point of quite hierarchy, and try to connect, as an independent work, this lecture, devoted to creating, for example, a union sphere of governance, which is stable anyway, and in any case you reach up the feasible global cataclysm does not happen. The case is that I always add the word "feasible" because the oral characteristic means it does not occur certainly. Examine this wording in inner content of sphere of governance as a task.

As the governance through self-recovery now here I examine a method when you perform the governance building yourself both from inside and in some one local segment of outer world, i.e. outside reality, at the same time. This self-recovery has a positive effect upon everyone - this is an indispensable requisite condition and it forms systems of the transmission of impulses in Consciousness, which form the reality in such a way, that Consciousness begins to work according to specific canon laws. In the giving method in the giving case I enter the system of the growth of Consciousness, i.e. increasing of characteristics of Consciousness on making special spheres which are canonized for everyone; this is a system of knowledge transmission.

And in this connection here I examine those positions, that examining an outer impulse of governance from the outer reality, you can divide it into elements of outer reality when they are manifested from outer sources: this is someone's information or from somewhere, or you have made the information by yourself that is not on principle in this case. Here there is a drastic difference from the first part where the question was about the governance of feasible global cataclysm, that is not on principle in this case: you may either fix the outer source, or to take any one from outer reality and in addition to examine yourself and within this to begin construction through self-recovery of the whole reality from two levels of perception. The case is that it is clear logically why in the case of self- recovery it does not matter: because as I have already said in the first part on the technology of the prevention of feasible global cataclysm - with the point of eternal development, a human form is ideal, so all outer events on the Creator's mind are human's action and the Creator's action as well. So, the whole outer event, for example, which you examine as an outer element, finally is all the same your own participation in the system of development.

The question naturally occurs: what is the difference if we examine the feasible global cataclysm, do not we?... But the case is that here feasible global cataclysm which may happen is not foreseen by elements of constructive development. So in that case when I told about technologies of giving for elements of reality exactly outer characteristics or reproduced ones personally by you, it had a principal meaning. In the case of the reality is constructed through self-recovery, it has the same meaning.

In this connection you have to examine independently such elements of Consciousness which are universal: a sphere of Consciousness works in some cases, in the case of outer reality and your appointed informative construction; and in some case the same sphere of Consciousness works in such a way that in this sphere the outer and inner are the same. As a result you realize at once in what way to make the governance on self-recovery, as you are in one place and in one form. That is if simply to consider that element of Consciousness with the point of eternal development, you realize you can always construct harmonious reality. As an example to construct worlds and develop informative systems only because you understand the structure of Consciousness. That means the giving principle of self recovery and construction of reality through self-recovery is based on that you construct your own reality in such a way that you are, for example, the first manifestation.

Then the giving construction is so that you can construct not only your own reality, your own body, suppose, and within this to develop in a harmonious way everything from outside; and this is enough to understand the Creator's tasks and plus your own tasks, for example, the tasks of everyone and you can construct the next element of reality. Techniques can be constructed in such a way, for example, which cannot destroy in future, but have the forms which are necessary for future development. So I used the word "forms" in its mechanical meaning here. The elements of development exist when even forms of techniques, such formal signs as: parameters, geometrical parameters, which determine the elements of eternal development.

If to consider the histories of civilizations, which had quite long development, in some civilizations in due time, in the past, - I mean the history of the Earth - the meaning of form was given a special meaning with the point of opportunity to be revive to life, but with the condition that certain problems may occur in the period of development. That is they founded a system of stability, formal, symbolic signs, and, generally speaking, it certainly helped the Earth was not destroyed now.

So when you consider the elements of development, be more attentive to the forms you create in the elements of cognition. Pay your attention here that in the elements of cognition you create forms. When you cognize something - this information stays in the kind of new, supplemented forms in the elements of the World development. These forms also can be controlled and governed. As a result you will not have certain out of controlled spheres of governance, and exactly this sign, this principle means that you can perform the governance through creation of your own form not only, for example, in material realization, as a human body or proper events, but and form with the point of eternal future development. I do not reveal this part entirely on purpose, as it needs to be thought till the end on the signs of self-recovery and to dictate mentally so that it is possible to dictate something else till the end according to the system of governance by self-recovery.

In that part which is the next, where the governance of events of any order from your Consciousness, I state basis on that part which I have not told yet and on which I make the governance of events from my own Consciousness. That means in the giving case you get the governance from the sphere, which is not defined in your Consciousness at least now, or if someone has already thought about one of variants, in principle everybody here in the auditory has already thought about; another question is: any case I base this lecture on the foundation that the element of reality is not defined. In this connection you should get exact governance anyway, when starting point is indeterminate. So the giving technology of governance is special and constructed as technology of the development of Consciousness, but and including the governance of wide range of events, including the fact that governing only one particular event, you govern the whole system in a harmonious way. Suppose, if you cure you at the same time cure almost, will say, all participants of events or optimize for everyone at once. It is not difficult to understand why such a complex is possible using this technology, because the governance is not declared and it can be projected in any place. So the carrying governance is not declared, consequently in the case of eternal development anyway it is projected there where you want it. And the task of your Consciousness is just to give the status of eternity through initial impulse by purpose sign, by the aim of the governance.

If in the meaning of the development of situation it sounds, maybe, somehow complicatedly, try to consider the system here when you can percept it easy. So, you should make up your mind to percept the material easy, and study this part from this position. So specially make up your mind that there are no some particular problems here and you study this part of material from this position of Consciousness. In this manner I definite a sphere one more time here, which in fact is unknown and the governance enters in a purpose sphere that you want: it can be a medical treatment, any event you wish to obtain, and this governance, naturally, goes through prevention feasible macro cataclysm anyway. So any substructure goes through this system in the same way as DNA is structured and it executes this work all the time in respective of the fact if a man governs Consciousness specially or not. And so the system of governance in this case is like that the impulse of governance, which at first, is already realized, that means everyone participates in some discussion through Consciousness one way or another, needs to be constructed, but -within this the governance is constructed on that positions, where the initial characteristics do not exist.

Consequently, the logical task is very simple further and consists of defining at least some limited conditions, so that to govern from the level of unknown characteristics: from what it is necessary to proceed for to govern. So, the limited conditions are the environment which contains the unknown environment. And the case is that according to the system of common interconnections that every element of reality interacts -with each other in any case, it is clear that the neighboring environment is formed taking into account the interaction with unknown environment. As an example, when in my time I made device, based on crystals, and it, will say, dealt with or its main task was, it is still functions, to prevent feasible global cataclysm, so the aim of prevention in the case if we have unknown characteristics, the aim of the giving governance in crystals was the crystal included characteristics of outer environment for governing impulse. So any impulse, which occurred, in optics was like that his parameters were already known as the form of the crystal and their situation were so that they also considered any outer signal like a signal included all outer signs around itself.

Mathematically point of view it is not difficult, exactly in Mathematics. With the point of optics, optics of perception, for example, the processes of the same type as in crystal systems often happen there; in this connection when you have governance, this governance at first can be current, so you may have current governance taking into account that you, as I said, have the governance; the governance from the future may also happen, i.e. when you have some sphere of governance, but -within this the sphere comes from the future.

Being at home or while being at working place try to transfer as far as possible as an explanation the term "to have governance", that is when you are in a condition of governance, to a usual situation, which, maybe, you suppose is every day; so try as far as possible to adapt the term "to have governance" to yourself in this lecture, so if you are in such condition that you don't do special purpose of governance through Consciousness, examine: how the governance happens nevertheless if you don't do it.

And here when I consider the governance of outer events now, the principle of construction of the outer reality and inner reality at the same time is reflected in that you solve the problem of the process of reproduction the events. That means when the Creator made the reality, He solved the problem anyway. If you make the reality in the same way as the Creator does, and this is how it needs to be done, consequently, the next element of reality is the element, which is not marked maybe absolutely exactly, but you have to construct it nevertheless.

During about five minutes of the giving lecture I tried to describe limited conditions for the sphere which is unknown. In this connection try to listen to this part of the lecture specially maybe for several times to understand where is the moment you are displayed, for example, in the situation -with the sphere governance which is unknown. In such a way if to lay down the limited conditions as basic, it is clear that your perception, i.e. you percept this element, and it is so that you consider the sphere as an optical element by perception and you suppose this sphere is an unknown sphere from which it is necessary to govern; but it so happens that your perception is the element of limited conditions.

So the case is that the spiritual position on the giving governance exists, which consists of your ability not to percept optical level and the same principle of "limited conditions" also functions in this case - here you should realize it in such a way, so that to choose the governing sphere independently. And the logical, i.e. the principle of perception, exists, which you construct as your own perception of this sphere is already an informativity. So if you generally say that yes, there is a sphere, which characteristics are unknown, this is a manifestation of some reality. And, generally speaking, the aim in the giving situation is just to work on mechanical base of the giving approach. So, distinguishing this sphere for yourself, you reveal it somehow. So you may shape it, for example, suppose, in some kind of sphere; even to tum it to the kind of canon or anything else you like; or just to leave it as you percept - so is easier. As a rule it can be some image not cleat.

In this connection let's make such a combined practical lesson now. That is, for example, everyone has to create an image, for example, in front of me, in front of the table, which is unknown on the level of information, but anyway this image saves from feasible global cataclysm or recover, or governs all events at the same time - according to three positions; or... or you may add "and", if you wish. That is in this connection everyone creates such an image on information in front of me. It can be, suppose, an image been in one meter radius, for example, one meter- this is more compact; and then some image is in front of you now. The giving image is a collective image, which everyone has made. It accumulates in one place and it is so that we have, suppose, such a governance. And in this connection proceeding from that the giving governance is marked as the governance, which is in the aspect of your perception, you may consider concretely because you know membership of the auditory, you may look around as from clairvoyant point of view, or when you entered this auditory you saw everything with your eyes.

Look, what in the giving governance it means: Consciousness of everyone, the joint Consciousness and your own one? So, we have three positions again now. And distinguishing these three positions, you may construct the reality by parallel, in fact some kind of projected level of making the reality. There is a table, as an example. The table is like manifestation of collective Consciousness, but collective Consciousness of everyone. Nuclear bomb is nevertheless like manifestation of collective Consciousness, but the collective Consciousness, which was initially based by someone for example; and the collective Consciousness of others - those who did not want it explodes, for example, in the giving situation, - this collective Consciousness is not activated so strong that it -will not explode, "\\Till it?... And the aim of this training is: to distinguish that characteristic from the sphere, which is not determined, which solves all questions of governance anyway; and within this it solves the questions so that you have governance on the level of any event nevertheless, also the most particular or even smallest events. So in principle the aim consists of the governance of any event nevertheless.

And in this connection now you just construct this sphere one more time - as I told at first time, the sphere with indefinite characteristics, - in front of me, it is compact - this is your collective Consciousness for this auditory. The task is: you need Consciousness which comes from others, another part of collective Consciousness must be linked up with these aims here, for example, the universal salvation from feasible global cataclysm and system harmonious development. Everyone should attach the whole collective Consciousness, including Consciousness of the whole future development, and in this connection when you say: Consciousness of the whole future development, you have to go into the sphere of eternal development and to examine what is the collective Consciousness of everyone in a case of eternal and endless development. So the aim is to examine the whole eternity in further and how this Consciousness is displayed, including if it is necessary Consciousness of the past.

In this connection then we perceive the reality in such a way, the aim of construction in Consciousness in the giving situation is determining those characteristics, which mean these actions for Consciousness. So, the giving actions for Consciousness are that, at first, and this is very simple, you, see the sphere at once in optics of perception, which means this, i.e. for example, you may say at once that, for example, the endless development of everyone - the eternal development- is a specific segment of Consciousness, which is on a specific distance from you, and in the kind of optics it looks, for example, like a line, and that's it. So, the aim is not complicated actually.

Another question is when you replace the governance not only logically, but place the governance on the level of Spiritual development, the Spirit is like a basis, like a system, which also constructs the perception, in the giving situation is displayed as an element of universality, i.e. things around. As an example, you may distinguish some optical element in perception and if you did it, everything that surrounds optical element can be considered as Spiritual manifestation from personal Consciousness. That means the Spirit can be determined, And in this connection it is also possible to determine the presence of the Spirit or its activity and its governance; and then through determining you may be adapted to what your perception means when you have such task and the way you solve these tasks through perception. So the researching mechanism of Consciousness consists of defining which helps to find the truth through the next iteration: after determining what the Soul is in manifestation you have some construction which means the giving action. And then, in the next iteration, for example, you make it one more time and come to the Soul's truth: either like to the truth on intelligible level, or you are in the Soul; and within this you see as revelation what is really going on the level of construction of your Consciousness. Then it is so that you look through your Consciousness, but at the same time you see the way it is constructed by the Soul. This is the governance of the Soul, which happens on that level that you make this governance, and consider yourself in many positions. So this is the position of Consciousness, the position of the Soul, the position of what the Soul reproduces, including Consciousness and the Spirit; and the best way of orientation is a straight contact with the Creator. That means in the giving situation you are on the same level of contact with the Creator and all knowledge you percept is knowledge from the Creator. In this connection the technology of getting knowledge from the Creator consists of you may come positionally into condition, when the knowledge comes from the Creator specifically. And in this connection, for example, if you need to get the governance, the obligatory governance on salvation, for example, it may be realized any way, you should only set the positional characteristics of such governance.

Coming back one more time to the aim of construction of that reality, which is not defined, let's construct it together in front of me one more time. So, for example, this time the reality is formed more clear, it has absolutely legible optical characteristics and is situated somewhere roughly the floor and begins to grow up. In first moments it were the impulses mainly, which were roughly in the sphere I told about. In this connection it is clear according to three times that the aim of the development is to construct visually objected reality. What is the main point of objectivism of reality? It begins to become apparent on the level of perception, and when we talk about the Creator's aims, the objectivism of action, the canonization of action - how it needs to be done, consists of construction, for example, of a human body which must be produced eternally. So the aim of eternal human production is the aim of natural development of, for example, one man. It means each man must be eternal. In the giving situation I have done only three repetitions and found out the law of development of information.

Well so try to hold such measures independently, will say, either in a group, or individually and control as far as possible the way the growth of information is going on. But when you, for example, have already constructed this information, then the situation is so that to get the specific applied governing action from this element you should - as I told in this part of the lecture - define the level of eternity in reaction of this element on your specific events. But it looks like that some segment is taken from this sphere, which is signed as the element of eternity, and this segment gets into that sphere of information, which needs to be supplemented according to your aim of the governance, and it only supplements this one. That is in the giving situation you have to single out the aim of the governance as, for example, the governance is supplement optical element up to the level of governance.

In the case of diagnosis you single out the aim of the governance in kind of some optical value, in perception it may be, as an example, some informative light image or, for example, even the man, whom you can see in perception. And the task in the giving case is a treatment if this is the supplement of optic form, or output of the governing impulse in a specific sphere of a man, even physical, for example, in that sphere where you get recovery too.

Here the task is to decide in what way you should do it in the case of ideology of eternal development exactly? You should know how to do it independently, to lead out the governing level, but as a rule, the world is settled in such a way on the level of governance, that the next your action, which is directed on this impulse, is absolutely right.

So I conclude that to construct new governance in such a way that any action you do is the right action. And the canon, which determines that your actions are right, is that you are up to canon level. That is, as an example, the global cataclysm does not happen, and everything is going on more harmonious with the point of your governance.

Here the problem of alternative arises: what is considered to bea good choice for you? The development of your Soul, maybe, of your awn cognition, - the task is more important than, for example, some actions linked through Consciousness of substance governance as teleport transfers of substance, materialization. Cognition is often more significant element than action, as prevention feasible global cataclysm, for example, or disease is significant in the system of governance. Within this the task is not manifestation, but cognition. So, the practice of manifestation ... you had some practice and knew the way to act from the point of technology, working for governance on physical and information levels is required. Everything you may display on physical level, - you may do it. And generally speaking the main aim is the governance up to information level, such as not to permit some possible problems. And within this though information purposes are more important, still try to analyze correctly the governance on physical level as necessary.

As I have already come close to realization the governance technologically, I just enter an axiom now. It consists of your governance, which comes from your Consciousness, gets into the point of your own Consciousness. Anyway you do the governance in your Consciousness only because everything is under the governance within the limits of Consciousness. Hence it appears that the task of projection of system of governance on reality is that you should connect only Consciousness segment -with reality, which means community of all points: the point of initial impulse, from where the governance is realized by you, in the giving situation this sphere is unknown; so and the spheres of governance, i.e. descriptive elements, another element of Consciousness, it means that only it needs to be transferred to the reality by you. The reality in the giving situation is the human health or some event, this is the physical reality. And then you display the same in a shape of optics where it has to be done, single out one more time beside the main aim of governance, just mechanically, enter automatically this descriptive point in the sphere with such at most fast, superfast speed- and that's it.

Then we get that the sphere that was unknown was not definite, generally speaking, it was not necessary because we look only at its descriptive part, at another point of Consciousness. And another point of Consciousness is defined and described, and can be analyzed a dozen times ... and as much time as possible. Consequently, this construction of Consciousness which I have just considered, it is so that many ways really exist, but in the giving situation I have considered one of the ways when the reality is constructed according to impulsive nature, as I defined, placing infinite characteristics of reaching the aim on governance or finishing construction of the form into impulse. This is direct, kind of frontal principle of governance, the governance conditionally forehead, we may say so, when you reach the governance according to this technology any way, where the element of reality, from which you started as from starting point, is unknown, but the governance still is going on. And you may even have superfast governance here, as the distance from unknown sphere to the sphere of governance is also unknown, it may be at most small and placed in the very sphere of governance. So really this principle gives superfast governance. For firmness, duplication this method, in fact, simultaneous, 'Will say level of realization - be attentive here - try to consider "simultaneous level of realization" as an independent homework, what sense I put in the statement or speak with me through telepathy. After the giving lecture the next part of lecture but in telepathy, will be in fact during three days from ten till eleven p.m. There you may ask the questions, connected with what sense I put in this notion.

Here I only examine the common point. That means you may only put the point of reflection - and that's all. And it is so that the sphere with unknown, the sphere with known is reflected into the point- into the sphere you just appoint. You transfer this sphere -I repeat myself again - with very speed, super high, into the sphere of reality, in fact, also designated in Consciousness, - you have just singled out it - the governance is obtained. This is the governance from the sphere with unknown characteristics in such a way in the giving situation. Such construction is cognized very quickly in definite situation, so, the giving governance becomes clear and very simple immediately. As far as possible do not simplify the giving governance and those complicated segments, which are displayed, - what the Spiritual development means, - come back to them again. Do not simplify the development of your own level of cognition in this part as far as possible, though I have made it at most simple, and it is really so; if you want to consider some segment up to the level of analysis - do it. Here is a special explanation for this part of the lecture.

Well, I have just finished that part that concerned the part based on the lecture. As far as possible to make in front of me the projection in governance from unknown part one more time: it has already formed absolutely exactly in the auditory and gives already the system of governance. In such a way now I set to that I begin to answer your questions. In the giving case the answers 'Will form the development of giving lecture, but not directly, surely. That is though the answers need to be considered as continuation of thematic from the view of set methods; nevertheless I will also introduce other methods of governance according to the answers. So here we have that the part based on the lecture is defining so that you understood at once when you work with practice in further that, for example, the things I say now may be used as work with some specific situations in fact. In such a way the first answer, well, at first, will concern the information itself, but in fact in one applied meaning - though I will also try to discover a new fundamental meaning of governance here - from the view of those methods, which were stated in the part based on the lecture.

The question: so, the question is: how to work with negative information?

The answer: the case is that the notion "to work with negative information" :first of all is so that here it is necessary to define at once as far as possible the element so that not to fix negativism of information.

This is the principle aim of governance. Consequently, the most simple way, in fact, from the view of methodology, suppose, governance with double impulse nature so that to light in Consciousness these two sources - inside of you and outside - in such a way so that they either light the meaning of negative information that it disappeared at most at once, at first through up to the level of perception; or to make maximum quantity of interactions of these two segments which restructure the information. That means to set such movement through Consciousness by rearranging Consciousness to automatism, that just to set the aim program sense in these two sources, and they are to restructure the information. The question: so, the question is: how to work with negative information?

The answer: the case is that the notion "to work with negative information" first of all is so that here it is necessary to define at once as far as possible the element so that not to fix negativism of information. This is the principle aim of governance. Consequently, the most simple way, in fact, from the view of methodology, suppose, governance with double impulse nature so that to light in Consciousness these two sources - inside of you and outside - in such a way so that they either light the meaning of negative information that it disappeared at most at once, at first through up to the level of perception; or to make maximum quantity of interactions of these two segments which restructure the information.

That means to set such movement through Consciousness by rearranging Consciousness to automatism, that just to set the aim program sense in these two sources, and they are to restructure the information. The question: so, the question is: how to work with negative information?

The answer: the case is that the notion "to work with negative information" first of all is so that here it is necessary to define at once as far as possible the element so that not to fix negativism of information. This is the principle aim of governance. Consequently, the most simple way, in fact, from the view of methodology, suppose, governance with double impulse nature so that to light in Consciousness these two sources - inside of you and outside - in such a way so that they either light the meaning of negative information that it disappeared at most at once, at first through up to the level of perception; or to make maximum quantity of interactions of these two segments which restructure the information.

That means to set such movement through Consciousness by rearranging Consciousness to automatism, that just to set the aim program sense in these two sources, and they are to restructure the information.

The question: and here is the continuation of the question - what means to break up the information?

The answer: consists of , for example, existing a form of information , a certain form, which, generally speaking, might be not fixed, that means grasped as simple strict form, realized into the future events. The main is it cannot be allowed and broken up further, only mechanical changing of the form is tolerated, so breaking down a form is another type of information, isn't it? ... - owing to, for example, two sources. Or else to enter the spiritual sense of Consciousness, which breaks up the form by itself. So, the point of cognition is put as the governing point in the giving form. As an example, I see a Dictaphone and I wish the giving information did not contain any negative. Then I try to go deep into the system: what I need to know if I look at the Dictaphone to reach breaking up certain negative information? Then I, suppose, should know for example, that photosynthesis can be in the exclusive system, so light presence is not obliged there. How to make it, right?... well, I find how to make it, and then I see that negative information has broken up; so I used indirect method - the method of cognition. What kind of negativism may exist with this Dictaphone? Negativism is not absolutely correct notion in this case, I just wanted so that the base I put in the purpose of cognition and with the speed of cognition on universal salvation, exactly that speed was got there. That is I have just done it, for example, haven't I?... but I could make it in an hour after lecture, to make one general impulse so that to solve also the same problem in real time. That is why ...though in real time I solve this problem anyway. Another question is: if to control all information when you replace, make the governance, you nevertheless replace and break up not only that negativism which you percept, for example, but you may also participate in all elements of reality at the same time and within this to get the governance only of yourself.

The question: the next part of the question: does this replacement of negative information mean positive?

The answer: the case is that in fact from logical point of view positive, yes - the form works in the same place as positive system. The question can be considered like it may be replacing, i.e. in your consciousness you may percept it like replacing; but on the other side you may, even without concerning the form, as I said, to consider that element of your Consciousness which contains any sensitive side on the aim of governance. So, without actions, being near yourself you may work from only one point within this, then it is so that this is not replacing, but reformation a sensitive level which you put into the governance.

The question: now the second question is: how to construct the structure of a child with regard to his will expression?

The answer: to my mind the case is that first of all it is necessary to construct the structure of a child in such a way that at first you should give absolute freedom status of personality in future, and also in presence or past. That is to :find out personal freedom in the conditions that the child is obliged to do, for example, what your want or socially necessary duties; and within this according to what is written here: with regard to his will expression - what he wants. And it is so that in this case the structure is to be constructed in the next way with the help of my technology of governance through Consciousness, structurization of Consciousness that at first you singled out the light, i.e. the optics of absolute freedom of your will, this is the idolization of child's action - i.e. super creative way of development. So you single out this information, then enter the information of social norms as, suppose, certain contour which conically closes the initial superstructure; and then enter the system of will expression as an inner system, right?... inside of cone, and in addition to this your own educational tasks.

It so happens that the governance is only tracking of this form. If the cone begins to change somewhere, there you see, suppose, that your child's will is suppressed somewhere, or he feels he is not free somewhere, just add some governance therein. But within this you may come into there absolutely specifically - to see what reduces his freedom: there is the unpleasant smell or maybe some actions of his parents or someone infringes upon his interests and out of doors? Just someone's look, which, for example, infringes upon him, right?...the freedom of his development, maybe a reason, not only someone's effect.

Then the task in the constructive context is not to cause any problems to anyone; anyway it is impossible to do harm using this technology - you just enter the information about compensation of giving systems in his perception, so if someone's look is unpleasant for him, than you add and enter the form so that it does not influence his growth, right?... Within this you also make good to that man too.

Why such a system cannot cause any harm?- Because at first, here there is a purpose level - the universal eternal development, and technologic optics does not cause any problems.

The optics is, the optics is weak in absent-minded level of Consciousness, and it does not across, at first, any other one; there are such elements, when you look at this system of governance more attentively, than that optics is, at first, non- linear; the light falls not directly though it does not cause any harm too - usual light, for example. But that optics is relieved, so the optical elements at first, may be far from each other; and at second, you reproduce them by your own Consciousness; and Consciousness is the sense of exact reproduction, when you make the governance exactly only in that case if you do common constructive actions.

So there a lot of duplicate levels, when you, for example, supplement that man's look with, suppose, correlated shade, right?... you do him no harm; on the contrary, you even maybe - "look" because he has ache of something, suppose, than you recover him and so on.

In this connection ... theoretically you may come deeply into the system of that man or - taking into consideration the things he did not like, the smell for example, - reconstruct the smell and so on in this case. So, with the point of this technology I think, as I paid yow: attention at the same time, the basis is, well, the main is to fix the freedom of personal development as the first element of action. The question: and now, so to say, sub-question here: how to construct the future child's structure?

The answer: with the point of the structure of information I have already considered the question; but with the point of the sub question, so to say, structurization can be particular, that is when you look at every event if you need it, but the main point in this case is so that giving freedom in child's development is an important action; that is to use only spiritual level of governance, for instance, when the freedom of his will is not broken; and within this you do not control every particular action through, for instance, clairvoyance level of approach, tight... ? So, each man is to be also independent and herein you give him an opportunity to act without assistance so that there is no control his behavior structure when some events happen, well, but you know on the Spiritual level of contact that everything is fine. So the aim of structurization of the child's future is, so to say, giving a free hand.

I suppose there is no necessity, well, to control physical actions, transfers for instance, the main is always to see the information on Spiritual level that absolutely everything is fine, that is to find such a Spiritual level of contact. Later on, evidently, if you think you need to consider the eventual part, i.e. to structure future according to eventual part, for instance, sit for examination, or some social events, ot something very specific; then, surely, you may work out in detail all events up to the maximum level that you need: at once if up to cellular or connection level including all connections with the eternity, with all outer and inner array - no limits here. The main point is that you set yourself as an object to get the universal constructive development, and within this the applied particular aim in the case of structurization of the child's future - to my mind there is an important point here, which emphasizes that only performing for a child, generally speaking, for all children who yet develop on social level - performing the particular task of governance for him you have already known that it is not necessary often to take into consideration macroatrays of governance because you already realize this task, because you govern for those who is not developed socially entirely, isn't he?... or physically. So the governance is for those who are weak, sort of weak, right?... so it always covers the level of macroregulation. And in this connection why, for instance,.. such a diverse opinion exists that if, for instance, to consider availability of super-developed civilizations, but they do not injure any harm for those, who, for instance, are not developed yet.

The case is that the common social principle or such common law of development consists of realization of governance through macro array. The following knowledge is so that the weaker system cannot be restructured, that as like the knowledge of the eternal development. It is so that macro governance realizes on the principle of governance for a child. So you may work it out in detail somehow and see that the principle of governance for a child is actually the principle of the eternal development; as this is enough to give a system development, to create the child's development on cannons, and then the eternal future is a stable process. And in practice... well, for instance, it happened so in my practice that children were presented at my lectures, herein they were also often sleeping, it means that they were mastering the material very quickly, than practically, well for instance, adults, herein solving other problems, maybe they had more deal with macrosystems. And in this connection now I think it is very important here that if you give the system of governance to the child, then the main poiot is to understand further that you put the elements of the eternal development in your governing impulse that is very important in the case of the child. And if to observe the way the child's development is going on when you already give the system of governance from Consciousness, here to my mind it is important that now further you may give by yourself the aim, such a special part.

And by the way here is my pupil in the auditory - Andrew Vorobev. It is so with him that his daughter was presented at my lectures not so often. And within this she could after a definite level of lectures, she was not listened to the whole lecture, for instance, uninterruptedly; she did absolutely professionally - began to make materialization. That is I, well... and when I controlled and made these figure methods of analysis, it was so there that the professionalism, for instance, consisted of the structure of governance was so that as if it was necessary to study almost the whole system through the system of materialization from beginning up to that level when it always and regularly happens.

So, child's perception is on the level of contact with information of reality on such a level that... he obtains this whole complex at once. Here it was of importance that genetically there was simply personal information in the development. But nevertheless the common law based on canon has an effect here, that, for example, to my mind, if to structure the child's future, it is necessary to take into consideration that any case children communicate with the more direct reality. And so it can be structured enough quickly, so to say, and it may be gone in often ... or constantly, or even often is enough, for instance, just to set the information you want -it is realized very quickly. So I have such an answer on this question. According to how much the materialization or treatment must be done there, right?.. what to teach a child, for instance. With the point of macroregulation there is one more detail that a child is more close to outer information, inner, so to say, reality, governing information; that the governance from the child coming through materialization always concerns With macroregulation. Because the child crosses obligatory this level of macroregulation to realize his future events. It is based genetically up to a certain level of child's growth. It happens always, because the child always refers the future element of development to himself through the governance. So it turns out that what for an adult, for instance, should contains, well, logical, for instance, status, for a child it contains natural status as well as own governance. So, here is the answer to those two questions.

The question: the next question, well: tell more detail about the methods of work with situations. The answer: well, in this case, I can tell you as concrete application, for example, from the three-volume edition "The practice of governance. The way of salvation" about the work through Tu-144 aircraft. There were test flights, and you \Vill just see as an analysis concrete evidence of Boris Ivanovich Veremeya, who is honored test pilot. There is written that I gave a forecast on the failure of the pitch attitude angle, and so ... the experts then found that the plane would have crashed whilst landing if such a forecast wasn't made.

The point is that here the failure ... when I worked ... now take a closer look at the specific situation, for example. The plane is to have the same optical scope, suppose, that is in terms of foresight activities in the future. So I construct an arc in the perception that comes as the starting point of the take-off of aircraft and, therefore, gets to the point of governance that, well, already is landing. So there's the arch I see the whole spectrum of events relating to the aircraft. Then I consider what kind of, let's say, nuances, of changes from the norm are actually occur near the arc. The f1rst change immediately, which in principle could be brought about is that the engine has been replaced, delivered from another aircraft, Tu-160 and, accordingly, it was immediately clear that there are questions already. And then I've considered, that the pitch attitude angle, that is, in principle, the directional angle will be unknown due to the fact that the device fails.

When a pilot takes a decision, he takes it based on the instrument data, and this information is immediately visible somewhere in the fust part of the arc, as he will be taking ... Here's a feature that it would seem to be an arc can be mechanically split into several parts and say that in the beginning is the initial period of the flight, and in the end is like the ending period of flight. Actually that's a particular practice in this arc; approximately the first third of the distance there was information about pilot's taking a decision. Why was it there? Because I'm looking for this defect even before takeoff. That is, it turns out the information of future, well as a sticky tape, is pulled - if I'm there pull by chewing gum, it is pulled where there is another anchor point. That is my presence, my search pulled this information closer, though logically it should farther. And it happened that . . . here it was clear at once that ... well this is I considered just as some things I see as just a plot to consider psychology of the pilot's behavior. Because if this is real - degree, such as this device showed, then they had to maintain the speed and even lower; if not - then they should have raised the speed from rating above, even higher than in the technical conditions of aircraft Tu-144 maintenance because the engine was from another aircraft and the failure could be such that, then just simply really they have such a seat angle, and perhaps -as that's decides the pilot- that maybe the device is not refused, and if the seat angle is that then it must slow speed, it obviously would have crashed. But as it, well, the crew was basically there, they wouldn't have had time to discuss the situation whilst landing, so it turns out that in this case in this the third of the fust arc, so well, in fact, as this segment of the part of scope, and it is in the first third I enter inside in Consciousness, just I enter inside this concrete situation by Consciousness and consider how to act here. That is then quite clearly, as a certain deviation that is near this part of the arc the concrete optical component appears, which does not apply to standard, which should be immediately revealed by task of salvation.

You'll meet in practice, that is, generally speaking, revealed immediately right there, and there, you can immediately see that this relates specifically to a concrete device; there is no problem to find it further, because that's enough to enter the information of the whole aircraft, and immediately you can see where this device is.

If you follow the development, for example, of information as a system of unpacked archives, I just simply entered there, and unpacked the next, next, next - and doing this found this device which was specifically called as "FAA", I think it was pitch attitude angle. Hence the ... or else there is a device called ... in principle, there is also a system of other devices. I also pointed to one which they just took down; it's like a trifle, which did not have at that the journalist, he's just from a side, from some newspaper - exactly about Kozloduevskaya station. That is, it turns out, the first manifestation of reality, yes? ... Then, reality exists as an external, i.e. an invitation of the Bulgarian national television is the element of reality. The second is a question of a journalist which, incidentally, was published later in many newspapers, that I said that there were no problems for two years, in addition I gave some really diagnosis which has been confirmed. There was a representative from the nuclear power station- and that's all, and it seems like this question was disallowed well. I said that ... they are very concerned about the issue of general problematic, about the feasible cataclysm of the Chernobyl type, and they wanted just like to close it. But, about two years passed, somewhere in the ninety-seventh ..a relationship, but it could be then principle. In general, in diagnostic of technique it is necessary to make decisions on the merits in this case simply. This principle, it is no different from that, for example, you diagnose any situation at all. So this principle, for example, the same arc, which I proposed, it realized there to the machine there or to own movement, the movement of any other person. That is enough to close two points in the plane, and consider the spatial area. It is rather quickly for perception to examine differences and transfer in 3D level, to work in 3D projection. Dimensional projection is commonly used when you want to materialize a card as ... or do so that cataclysms -will not happen on governance at a distance.

So in this case, as they knew about the pitch attitude angle, they simply increased the speed from rating there, in my opinion, at twenty above, and it turned out that they do not split, though when they ... landing was there nevertheless next to, empennage is almost back to near the track still was ... That is, it was the most minimal level which made it possible to maintain the pilots, including also this machine.

So, the case is that the concrete methods, they are often defined by situations well, and one of the situations for example here about Kozloduevskaya station, this is when the macro cataclysm of nuclear power station could occur - the governance was formed in fact well from two periods there. This is the first - when I came at the invitation of the Bulgarian national television in Bulgaria in uinety-fifth year; and the fttst question that the journalist had asked- 30 © Grabovoi G.P., 2001-2003 year, somewhere in a month again information revealed that it I necessary take actions. But taking into account that there were a lot of work, this is just a back-to-hack somewhere three days before I did governance, transmitted the information, and then this is how it was published in the Russian newspaper - they were inspecting this information; i.e. they thought about what was a result; it turns out that if the actions have not been taken there and diagnostic would not have been given, for example, the explosion would be well this has been a problem from the point of view of feasible macro cataclysm. Therefore, taking into account that there was the situation, concerning the development of planet, in the problematic of explosion, because that means the situations connected with these levels of feasible mucrocataclysm, there is a future reality, it gives governing index in more intense form. That is salvation ... why I give in lectures that salvation is always possible? Because outer reality very clearly defines the characteristics of action there - that the planet can tum into dust, that is, fall down, as it turned out about the year 2000 because of this fistula, this reality, it is very clearly focused specific corridors through governance, here's how to do now that it will not happen.

There uniformity of actions arose in principle, that is absolutely precise level of governance, which had no other variants. So here, in principle, there is governance of the real systems of salvation, that is, when you really save, it is fundamentally different from the system, for example, governance, well connected with some hypothetical experiences or experiments, and, as a general rule, if you look through my materials, I just simply do routine work.

Well, it is kind of like experiments in aviation, although it's a real rescue of aircraft or cure, and so on. That is, .in practice of governance I give only as a very concrete technologies of governance of only real processes, and therefore in the future, you also better have to try not to contact there, let's say, with some variants of the testing, the repeat, because the goal governance is to achieve a real salvation, notably from systems including an unknown nature. Through repetitions, this cannot be• confttmed. By means of consistency: the main aim is such a spiritual condition, such an opportunity of mind, such a level of soul, which can in any case be saved; this is a technology, plus concrete practice through real salvation and, in fact, just through the real level of life.

Why that's the way, I still always transfer all the time through governance of own events as far as possible, including particular, because it is a distinctive practice that allows you to achieve great results, well, to gain experience. And then you will be more prepared than if you just occupy yourself with, for example, systems of salvation from feasible cataclysms, and in the same time you think that your everyday life is something else. In fact you govern anyway, so I try necessarily to transfer governance to the particular one. Of course, another question is: how quickly you get your governance? The tasks of Souls of the most are higher than their specific tasks, and then they pretty much do at the macrolevel.

Here I have a lot of students doing macrogovernance well, herewith and often quickly they reach concrete governing systems in the particular tasks. But, for example, when interacting with them when I analyze some situation, I see that it is not often their objective to do something quickly; but it's generally a normal thing. But nevertheless, I still give the governance through particular tasks for keeping you constantly in practice. After a while you get used to it, and then it will be naturally for you to control all reality on the level of at least on level.

At fttst, it may be somewhat tense situation; this is sometimes the first time in years, delayed until the person does not reach the level of constant control and at the same time, he already quite clearly knows that his system of governance can easily prevent, well, the feasible global cataclysm - that's exactly. And I think the meaning nevertheless exactly is to increase simply the amount of practice- this is the main meaning. The question: so, now the next question: is it possible to restore a deleted surgically, in quotes, organ?

The answer: the fact of the matter is that the meaning of governance is to govern to the extent to which you need. And giving the status of universal salvation of, for example, restoring the organ yourself, you can pass on this technology to everyone, and this is, of course, I believe, can and should be done even if all the more it has consequences on surrounding tissues. And in the ideology the goal of governance is complete restoration of tissue, including the fact that any instrument systems that do it, they only do the work actually of upsurge of the collective Consciousness through learning; but I think the goal is to do based on Consciousness of the man himself.

The question: the second question: it means the value of the Christian Church today. The answer: Well, I believe that any canonical religion, for example what is displayed as the Christian Church, where there are such commandments as You shall not murder, and so on, they always have, I think, a principle basic meaning. And so in any environment there, in all systems development, I believe that their meaning does not decrease, and in the future it can increase naturally due to the fact that the principle here is implemented, how the canonical principle, such as the principle of "You shall not murder", the principle must be the more capacious meaning in conditions of the system, developing system, for example, of global destruction.

And therefore, I think that it is necessary largely as a supporting foundation to use those things that are accumulated, for example, in the Christian Church from the point of sustainable development. And at the same time, I believe that we must bear in mind that any canonical religion, well such as true canonical ... today many canonical religions are known, canonized religions, which in principle set the same task as the main task: human reproduction, the eternal human reproduction. But there, of course, they are distributed there by regions, by representatives of the believers; and yet, the Christian Church, by the way, here displayed the question of eternal life evidentially. In connection -with this here, I believe there is some unifying meaning of the Christian Church.

So the point is that, if we consider the principle of further development as the element of eternal development of information, nevertheless the uniformity in opinions or uniformity in religious manifestations in respect of eternal development displays.

So everyone -will have his own view, but it means uniformity in terms of such laws as, for example, the force of gravity is the law; and then in this case it is the collective law for all. Therefore, I think that, I believe, .it is necessary to support generally any canonized directions directed to creative development. Not only the church, but generally any social movements, individual manifestations and so on. And herewith try as far as possible, apparently to help as much as possible, because the sense is that, for example, achieving the salvation of everybody and in any case is the basic element I put in my technology. Then it turns out that we don't have at all any particular manifestations really anywhere: neither in concrete person nor in concrete state for example, everybody should be saved, it means that everybody should have the technology.

In terms of why, for example, there is exactly such a process associated with possible there and occurring destructions there, various demolitions; and there is no, suppose, such an action that everybody instantly there saved, and more like doing nothing? In my opinion, the answer is that it was exactly the intent of the Creator with the full freedom of the will, if there is no full freedom of will, it is a unstable system, generally speaking, in development, the intention of the Creator is with full freedom to provide an opportunity for stable development, and it is only knowledge.

Here is, in my opinion, the destruction there, elements of demolition - I consider they as still lack the knowledge to a greater extent. And therefore, if we consider how we can construct further development, generally speaking, on principle and for what? You must construct, in my opinion, on the basis of knowledge only. Here is what for? To all the same it was eternal. Because the main sense is still the eternal development. If we're talking about the governance, it makes sense, in my opinion, to speak only about such a governance, which leads to eternal development.

Otherwise, we use only some evaluation systems, and exactly accuracy in eternal development gives consistency. That is, suppose there's lack of weapon systems which could destroy. Special protection technologies, self-defense through Consciousness or through special systems which cannot destroy the Creator, and so on. I have got just a number of works connected with principles of construction of technique which does not destroy the creator, and - well, it's more monographic works - but I think that soon I'll be publishing there a series of special monographs, you'll be able to look through the technique.

The question: the next question; -w hat is a generalized image? The answer: the thing is that now I'm reading the questions selectively and choose the questions that give the governance here closer to the topic of this lecture, I am just so represent them Generalized image, if we put some tasks of governance, then this might be an image that includes a number of systems related to concrete event; for instance the event is: I walked out of the conference auditory and there got into a car and drove, - for example, there are three actions, huh? ... called sentences for convenience: I went out, got into a car, drove there. The generalized image of this event is three elements of the event. That is, if every form is one of the forms. Here's how we construct now the generalized form, in fact, when construct an unknown element from which it is necessary to govern. It turns out that exactly the image is generalized - if three forms, if they are displaced somehow or put together, it's not necessary to put together mechanically; it is possible that the contact of these three forms or infinite set of forms, or some concrete forms will result in a completely different form, or even another manifestation of Spiritual knowledge.

There is a principle of work with information where you can obtain information only at the expense of generalization of informative image. Generalization means, respectively, the result So you look at the object of information when there is a task of generalization. Then it turns out that you see a generalized image as image which, firstly, exists and in those part which is unknown. That is how you can generalize the image? That is to take down some concrete characteristics and for this end it is enough to have even a concrete image and add to the image a governance, which we have constructed with you. So this image where characteristics are unknown, then you get a generalized image; but a generalized image is a more stable image from viewpoint of governance.

So, it turns out, added even what we didn't define, we nevertheless... well then we have a chance to receive more stable sphere nevertheless; and this more stable sphere - it is more governed. Because the more generalizations, the more possibilities to govern the system, there are more signs and more elements of entry in the system. So you should consider generalized image in such a way from the point of view of today's lectures.

The question: the second question, the next: so, the characteristics of the governed influence when working to prevent feasible cataclysms, self-recovery, governance of events ... in brackets.

The answer: I specifically do so that this is clear to understand what is the meaning of the question; and I give generalized for those who are in the auditory, individually for each person, in the manner that in the infinite future the answer , have been understood. That is, I immediately construct at least three constructions of governance, as far as possible in further future use immediately the same position of governance when answering questions for those whom you train or simply with whom you communicate.

The question: in brackets is written: indication, command, request, desire, exact indication, points of application. The answer: the fact of the matter is that characteristics are just yet those elements expressed as words which might be changed not only in what indicated in brackets. That is, it may be in principle any verbal construction, especially to prevent, for example, possible disasters. The thing is that the Word can be found, on the basis of the information. So if you want to use a word for governing ... I mostly give governance through the perception of optical elements, because here it is easier, often faster to achieve governance without words, because the optical element gives such a Spiritual condition, which realizing, gives the governance, but if you look at something, you have created the impression. Similarly I construct the governance approximately on such a basis, shall we say, like almost, when I, for example, give the images, and then you have just the Spiritual governance of that you are simply in a Spiritual State; and there is no need to then construct even optical images- you're a governor in this moment. If you construct a word, the word is made up of a series of optical, for example, segments, eh? ... or from one, but usually the word is often consists of many optical segments, because the word brings a sense, after sense also optics and so on. So it turns out that there is some superstructure from the top, and then the whole periphery of effects, which says that the word is used or here or there, and how it generally looks in terms of optics.

But nevertheless, I always admit that a very long way for the governance is possible - as a sequence of words. There were four hundred sixteen words in ancient Tibet in order to cure there literally even one scratch. They even did it to, for example, by the way of mantra indication they read, i.e. there has not been too here what ... what there's something less significantly or, suppose, something more substantially.

And so the situation is that to find a word, huh? ... or five hundred words - as you want - and how to call, how to apprehend? Here are the characteristics of governing action if still through the words, although here it is clear that you can specify the indication as optics, and act through this; but it is better in the intermediate part not to construct directive optical systems, if I even put it as a instruction or as a command, yes? ... then I get another link. That is better to govern directly without contacting the definition of intermediate systems of governance. Note that I came to answer not at all. That is why, here's how to call - a precise instruction? a command? For you it is better no call, better to govern. 'When you look, for exaruple, at the table, you do not tell yourself: now I'm looking at the table, and then you are looking at the table. Here is the same, so you're just looking at - and that's all, or doing something and that's all. I have a principle of governance, it is so that is to do, but not to call if possible. So if you still call, as I believe that in the system of salvation this may be of principle, that is, if you think that it is necessary to make instruction - make it then. And then how to get to know what it must be - instruction or still precise instruction, or request, for exaruple, or even wish? That by the way here is typical: the issue here is properly constructed in terms of different positions, for example, until the request or wish.

Here the situation is that here the main thing is, in my opinion, to identify that optical system how is the word in this reality born, then it -will be necessary for you to see how do you construct the word for yourself. So here's to see the word, you need to consider first your personal system - this is an essential condition in obtaining the word's information. So, you consider yourself, consider the information of word per se, and the word must be considered necessarily in absolute projection, i.e. you should consider the word so that it would not depend on anything without connection. And this optical information ... it quickly flows into your own information, because you see the word; and then you see how the word starts to work actually on you. So in what element the word is displayed there? There is such a concentration of f1ngers: you can do just a word, eh?

... You see, where the word's information is, from what part of the body it starts to highlight. It is very easy - you just simply look at yourself, and you immediately see that word's information entered in you and starts to highlight. Then the word's organization would be the fact that, f1rst, you can immediately be aware of this word spiritually, here is the first element through the perception of already an output parameter in the word is the primary spiritual awareness of the word, the word, in information first always according to my technology has been aware of spiritually. So what is the difference of words, for example, from an image? Spiritual awareness exists preliminarily always, i.e. there, where there is no word as yet. Thus, the second element is the presence of Soul in the word, that is, what Spiritual, eh? ... Soulful meaning you put into the word? Then you should see already on optics and retrace the impulse of the Soul, that is from the Soul, which is, for example, within your physical body - it is a piece of the Soul - we're talking about whole system of information or about displayed piece of the Soul in the body - and the impulse goes all over. It is again concerned of a word which was not announced yet heretofore. That is, so far I talked about the spiritual part, then about the word and so on. Impulse of the Soul and the spiritual impulse, which you feel, because logically I lead you in such a way, for example. Although you can choose your coordinates of the governance of word, and in this case, I yet only lead to reproduction of the word. Therefore, these impulses lead you to the Creator. That's when you see the Creator- you see the word; that is why the situation here is that it turns out that the word is a path to the Creator, or the word from the Creator. That's when we have, for example, the word, we must immediately understand the meaning, which is put into it, and then it turns out that you can make again the instant governance -without inquiring into the system of connections of this word.

I suggest again anyway, that how to make instant governance, if the talking is about macro cataclysm, eh? ... possible. So that's why even seemingly intermediate construction revealed in the word, you can lead out in the system of instant governance -with such a reasoning. Then you are quite clearly get a whole sentence and then you can write out what you have done in the concrete situation - this will be simply technologies which specif1cally will act there, at that they may act for eternal time, the important thing is that the point here is to lay io them the necessary sense of further knowledge, because the meaning of eternal development is always willingness and possibility to govern anyway; and if there are even very good systems of governance, they need to be provided -with the construction on eternal governance. Here's the way I think appropriate, consider the information about this question exactly with the poiot of double impulsive governance when outside and inside you on the principle of the governance of reality from the very recreation, here, I think, be useful in this regard, consider such a characteristic that the word at once always has two meanings: that is, the meaning of carrying- what the word carries, eh? ... and carrying construction, the supporting structure of information and the meaning of information, in fact, the very word. You would consider this yourself as an independent work on this lecture.

So, when you create a generalized image two images are clearly: the second is far, like io the distance, and on the lumbar level, like white mist. The fact is that ... well about the two images io perception: there is a image and there is a carrying construction - yes, they may be flrst, separated. So I have just said about the word, that means there's a carrying construction, and therefore one image can be considered as a construction that contains, for example, target oriented informative image, and it is really so vague - the second, or maybe even as a single point. Using the constant practice through the governance the carrying construction is often just a point in the image, and here is a situation that you can through Consciousness in principle to define among other things the carry construction. Therefore, I thiok that the situation here is that, generally speaking, if you see in the governance two images or more, see as far as possible: what is the sense of such a governance, that is what you want to get one for yourself, that is, what your Consciousness and your Soul constructs, and for what purpose you see these two images.

In general it is necessary to have precise information as possible io regard to any image and within this, so well, ... how do you get this information? why do you have any images? why do you construct the governance exactly through those coordinates or through those systems of knowledge? If you want, you can ask questions on them, finding positions like as construct questions if you need. Just in the governance, there is approximately such a status that if you see by eyes - here be more attentive, now I am going to try on the spiritual level to give it, such a principle - when you pass, somewhere out there on the street; here I sit and read questions or give a lecture, I perceive elements of reality by my eyes or by my Consciousness and construct the governance, then on the governance, on the level of the development of Consciousness through the governing constant, there the same state often exists that everything is clear for you anyway. That is, for example, two images arose and it is clear for what purpose you need them without itemizing anything. By the way, this is a very important characteristic of the governance, connected with that if possible, do not slow down artificially, i.e. don't learn there, for example, such as mathematics or as a strict science requiring the repetition, my system of governance, unless you need it. You can learn if you need, but develop as far as possible such a level of governance, when you understand everything that is going on. That is, and when you, for example, continue to do the governance, it is naturally for you as I said there, do, for example, and I understand what happens when I just see by my eyes, and also it's clear for everyone who sees by eyes. This is an element of collective Consciousness: everybody understands the same things, and in terms of collective governance through my system of governance try to construct such tiers of information transmission. That is, if you, for example, do governance, to think of others as far as possible, because in this moment it is not a universal system. If you construct a system of governance through how to prevent a catastrophe, and you really have prevented it- make it clear for everyone.

Then you try to go to the analysis of the current reality, as I said, it's just simply considerably logical analysis, that if I understand there is certain hard systems - they are understandable, because we see them in the optical range: doors there, some corridors or even an external reality. Construct the same things in the system of transmission of your knowledge - what you understand is clear for others.

So the task of universal salvation is practically a task including the super-fast transmission of knowledge, but also the task of system, detailed, sometimes can be very long, but the main thing - the compulsory transmission of knowledge through those technologies of salvation which will be found by you personally, or you'll get to know that they were found by your partners. That is, with whom you communicate, with whom you work, your friends, acquaintances, strangers at all. That is, it can be anyone who ... maybe with whom you incidentally crossed somewhere in transport, but in their knowledge systems can exist, which may be inter-related ... well, may be generally transmissive. So the task of the general eternal development is a realization of the continuous system through the transmission of instant or system knowledge. It is instantly if everybody needs to concentrate, concentrate for avoiding catastrophe; and system knowledge is that when we say that anyone can construct a system of education or something like individually to transmit knowledge to other. Well, today I have replied all questions that now I had time to look. On questions that I didn't review, I will in any case individually transmit on telepathy for those who wrote the questions. And in principle further it is preferably to leave some contact information because some questions I answer so that I dictate on the recording tape and just give if I'll decide, that it is better. So, I finish for today, so thanks a lot for your attentive listening.

EXTRATERRESTRIAL. DOOMSDAY IS NOT GOING TO HAPPEN

Was created by Grigori P. Grabovoi in 2002

EXTRATERRESTRIAL

He was an extraterrestrial because he saw himself as if somebody else saw him. He apprehended his eyes being a reflection of the external world in whole. And his soul flickered •with light of distant reality which was eternal, it was his eternal future. He saw the same people as himself surrounded by that reality because when he has met them he has understood that they have not been distinguished from him. When going his own way he incurred his inner light toward himself and toward others simultaneously. He gave away the light so as others did not interpret it to be an outgiving. They perceived it as an action. When getting sight of their wide open eyes and reading their wonder stricken thoughts in reaction to his actions and movements he considered them being organized the same way and apparently they had to be the same whether he was the one which he presumed he was. When setting eyes on first girl he thought that life existed in distant worlds because he was a man. And looking at himself in relationship with a girl he understood that love was in him initially. He would come that way and would see it in a shape of a woman who might be his significant other.

Going the distance of his journey of life and getting sight of rivers' streams change, what was going on to plants, animals and looking at peoples' life he understood that all had to be changed with regard to eternity. And his comprehension of that kind was not his word or a thought but it was a movement of his body. Namely the body was moving towards the eternity. And just later he began to perceive this movement as life of other people. He understood they were composed of an eternity as well. They were not particles of eternity but the eternity all by itself, just towards deep space, out of sun limits or even inside sun itself were speaking that they were who they personated to be. Though a part of them only was conscious of that. And his movement was directed aside to open outer space for them to see where their life took place and that they had come from universe and that they were to perceive their soul extended towards outer space as unity of the entire and singularity of exclusive. Having seen what the rest had seen he told about what other people thought but told to himself exclusively first. Then he began to talk to very many people and people started to follow him. They began to believe him and many of them confided in him their innermost secrets that were intended for eternal development. And very many people told him: 'Yes, we see our eternal life. You tell us though how we could bring it to others'. And he told them. He gave them figures and the life which they have seen by their own. He did not name himself at that. He walked forward leaving the name only covered with birth date. They were the same persons. Their movement was not always steady. They were to overcome a resistance while those distant worlds and their after sounds which were perhaps just developed up to the level of eternity in its stable realized phase lead them all as united cohort to their endless future. And this wave of eternity engulfed all when joy appeared even in hardest in peoples' opinion situations. They tried to dignify him however he always walked with them. Because the way they go is the one and the sky and the earth, everything that surrounded them are creatures of among other things their minds. He taught to create and to construct and to become a part of eternity because just this eternity was renewed and strove to be oneself whether one or other which occurred inside this eternity had properties of eternity at all levels. And its separate movements as well as all people movements being local and expressed by some certain emotions and actions were united by one deep eternity soul structure. Eternity which has a soul is an embodiment of humanity and essentially a forming image and eternity level itself. Whether God made it he approved himself as far as this approval just exists for ever within the frames of nature beauty and within the frames of dreams on future eternal life. And just here seeing that he had knowledge on eternity and having been here and acting they went after him more stick together. Nothing like resistance of the regime or one or another property could not break down them. They got stronger and collected moral courage and felt that they were not alone in the universe and that there were a spirit what would open before them those worlds which would warm their hearts in far journeys towards exploration of outer space. And they were glad to be created. They were glad for the God who was able to create this so they were walking with God in their hearts. They were revering and their movement just did not experience an impediments. Yes, there were some nuances, same resistance and sort of failures from those people who did not yet walk this way. But these who did actually did not experience failures.

They saw nothing else but action. It is just as speedy vehicle which is gathering highway. And sound of wind infatuates a human soul and gut and his body is rushing with a vehicle move momentum, and all the rest is not comprehended as something that is capable to detain this vehicle, that may cumber this motion. A human begins to feel like moving. This advancing circumflex by means of thoughts or form of thoughts became such an envelope that made these people invincible. So much invincible they were that nothing could made them to lead astray from eternity way. All they were and stay now common humans. This fact just attracts next people to follow them. With due rime when all people will be united and -will be able to live forever it just remains to recollect those processes followed to.

Namely their memory and recollections about human breath completed loads of things Extraterrestrial. Doomsday is not going to happen Extraterrestrial. Doomsday is not going to happen included unbelievable from the common people senses viewpoint is of importance.

Their recollection gave them strength. Their memories and their culture forced them towards eternity. And the eternity \Vas opened in front of them. They saw that sort of virtual heaven gardens what they could feel physically. They were alive and also so called eternal life in heavens within physical body was opened to them. And they understood that it has been completed for very many in the earth who has perceived a nature beauty; for those who has seen how leaves' sound has inspired a memory of past and all future events as far as they have been able to foresee. They knew all events of the future and it could be called a memory with regard to what they have already accomplished in their future of their own eternity. And the eternity engulfed them who have continued this way and they survived.

DOOMSDAY IS NOT GOING TO HAPPEN

Verity of deliverance is inside the essentiality of each of us. We choose the way which we define by ourselves. Because the way which is chosen by us is our way indeed. And we are moving for deliverance towards the direction where a light of our and external sensorium sets. Deliverance external is just an inner salvation. We unite with our love everything around us. And when those close to us and the outside world see that love world become unconditionally saved. Therefore first way of salvation is the love. And either conversations, either information on end of the world are to be given over to the fact that it is no any doomsday because a light as well as life is endless so as development of the earth and of all the universe and world by itself are continuous. Essence of continuous by itself derives the world eternity, eternity of living and eternity of each person's life. Life of each person is enveloped with eternity whether he does not suspect when he touches it at the moment of his birth. If you consider a case of birth which you know you see that before a bearing to the world eternity touch with birth moment becomes the very first act of life being eventful source of birth. The act as first act in life which appeared initially from first birth in the world primarily carries into itself eternity of eternal life because eternity as it is divided to fact before consciousness of it and after the consciousness. And this transition between a level of consciousness of and a level after consciousness just is movement of life at knowledge. When it is conscious eternity that is just beheld it may not be any doomsday. Because light is an inner content which is sensed by a human being and is a domain of apprehension which in substance is enveloped within some alive reality and a person responds as it would be a messenger of eternity and eternal worlds. And any description is something that consists of light.

Therefore first of all a salvation system might be intended at apprehension of events happening around. Any external world's systems are considered to be stable when we are looking sort of through them. That is going someone perceives happening around events so as surrounding sets outside and this external reality is a part of eternity. For eternal development would not be retarded - with piece of information on doomsday you have to consider eternity volumes in all outside world whether you see surrounding systems and outside world's movements.

Expanding your physical movement in material world you have to try to see a sphere of eternity which has its origin from you to closest thing and spreads endlessly up. When you are looking above extremely you see a light going to you and being your conscience. You have to consider closed conscience-the world system and you will see that your conscience controls everything. Therefore change of the world in general scale that is called the end of the world cannot happen because one thing having been seen once has an infinite construction filling the entire world. It follows here from that your arms, legs and body and your thinking and your organism cells as well are focused on one the task of eternal life. Focus vector exists absolutely because it stays 'Within love and it comprises elements of love. Thais is it is may not be demolished. Therefore no events outside may impede for love vector in the field of information.

When you start to consider eternity as phenomenon perceived by your organism and by all your essence you see that your separate tissues are already eternal And whether to consider your organism according to them you see that it is eternal. Therefore you have to move towards consideration of entire reality according to your conscience from those elements that you perceive to be eternity of the organism. At the level of conscience these are separate tissues of organism for instance in arms, legs, other parts of a body what eradiate silver fluorescence laterally to eternity universal for all. So collisions of your fluorescence, fluorescence of your organism of actual body with eternity for everything, which is just the earth having infinite development. The place where a human being resides at for instance the earth or at some another place out of the earth, and all these systems are united at your consciousness level with desire of eternal living. As a matter of fact a human being is born by reason of eternal living. For understanding you are to consider entire phase of eternity which exists before birth and before of a human being conception and entire phase which comes after. It turns out that very strong will to live only results in a human being birth. Prom point of view of world logical processes consciousness logic it is absolutely unequivocal. It is the same as to overcome huge quantity of information and to carry out largest internal work to get realized within physical embodiment and to come into the world. Therefore a being begotten man always overcomes namely the same part of work and the same internal system for movement to eternity as a just living man who wants to live forever. Having in mind that the fact of a given man begetting took place you may understand that the man has just overcome that some gap in frames of information system before bearing and after. It is an act of spiritual negotiation of a human being personality of all whatsoever impediments for being born and to stay in physical body. Considering a consciousness as indissociable system of connections 'Within entire world you may find out this consciousness element at yourself. It may be found out at the level of hypothalamus of a human being brains, at the level of hypophysis and at the level of tissues of for instance arm and legs. Thai is whether to look through finger nail growth you may consider one type principles of evolution towards side of eternity. Namely original spiritual constituent that was leading a personality up to eternity direction stirs organism tissues towards eternity the same way.

That is entire movement happens in the direction of eternal evolution. If one pays attention on many cases of peoples' survival under abnormal conditions he often may not get logical explanation while exactly this impulse of body's movement up to eternity transforms entire situation to the position when a man becomes not exposed to any impacts. Thus it may be clear that arrangement for his ability not to be destroyed was set in a human being initially. This is obvious in the instant case just after comparison information before a man begetting and after his first coming to the world after physical conception act for a man. Having examined given consciousness constructions one may mark those consciousness constituents which have continuous connection already before the moment of begetting where arrangements explained as willed arrangements appeared to be unconditional. That is urge to eternal life is not simply logical striving but expression of entire essence of a human being coming from integrated God the Creator who is eternal and thus it is his will. Eternity of a man is God will and human being arranging structure. According to that one may consider abundantly many information structures, intellectual actions which as well lead to entire and essentially detailed understanding of eternal evolution process and allow developing of technologies of eternal evolution independently in practical terms of everyday life. Thus we can in practice of any everyday deed to see that for instance hand moving entails an eternity illumination and hold-up of moving, when you doing something, allows to see that this illumination spreads from the hand as well as from a body in whole. With the reference of that eternity illumination has a structure of such sort that it makes a human being to reproduce this illumination. As much given illumination effect spreads to external reality so more techniques of eternal evolution a man obtains. %ether at the strata of thought you would see that your thought would start to realize such an illumination of eternity you will master to increase volume of irradiated light up to external reality either. In order to solve it you need to direct your thoughts towards eternity and towards techniques of eternal evolution. As a result your action will subsequently start to bring a picture of future development which is just emanates from you and from your light up to eternity. That is how you pilot your certain way by which you are going. You are of your own an implementer of illumination within eternity and of road of eternity.

This is apparent for the case because of fact that as more illumination you radiate so more reliable and more stable your movement towards eternity and eternal evolution is. Your eternal life must not relate to any information which refers to either isolated incident of particular person from point of view of his existential structures' demolishing or mankind on the whole. Initially there are unique systems of reality because entire mankind and a man are worlds on the whole unique, untouchable and which may be developed forever. According to whether we are speaking about an end of the world relative to expressly physical demolishing of dwelling sphere, for instance of the planet where people are located, then above all there is necessary to focus spiritual actions of a man in person according to a principle of mirroring at ourselves such information systems which come laden with information on an end of the world and further to pass to mirroring from entire planet and from the world on the whole. Therefore work over salvation for all people are to be made by means of movement of ourselves whether you see a structure of eternal your own evaluation and your illumination of eternity which is reproduced by your mentality.

When you find out the field of information you understand that it is very strong technique of evaluation of eternity technologies of your own thinking usage. Exclusively your own thinking is one technique and cogitation connected to elements of external and internal reality is a second technique. Whether you use your own cogitation technique you have to consider in the first place what peculiarly has some elongation inside organism in thinking. It shaped like a cone and spreads along organism so as top of cone is located higher than a head. Further due to principle of increase cone-shaped image of perception at certain point starts to widen around and above at that and grips a man completely just by means of external events. Thus eternity structure defines also these events which happen around us in such a manner that you may learn rather properly to live eternally. It is comparable to whether a man is supplied with a piano he may master playing in certain time. If he simply touches keys one may say that he already uses the piano. The same way to use at eternal structure.

Just move to eternal structure means a contact and usage of eternity realization methods. Intricate mastering of some music systems by means of the same piano or other instruments as well as by means of artist's brush or through words and through acting of your own are essentially practices which are given to people for they being familiar with eternal life system and signs of eternal life in frames of their evolution and their desires. Except that, it might have been set very serious element of eternal evolution because simple actions as well bear eternity what is worth to notice in one's zeal like you may come to cafe to drink cup of coffee.

Each moment in life and each twinkling carries very strong element of becoming familiar with eternal evolution structure. \\!hen you start to see it you may train techniques of this evolution which for sure will save you if any catastrophe, cataclysm and problem of inner or outer cause may be. Therefore survival of your is mandatory. Eternal structure being set at a man's cognition is located at information system and at information communication system so as when a man get to know that he can solve certain problems of tactic close level and also frequently strategic problems he becomes confident about salvation of whoever else the same way. Two levels of cognition support that. They are perceived as illumination coming from left and right half spheres of human brains. Illumination is considering as first impulse coming from brains increases gradually due to entire body illumination. We see that a human being's image is indestructible. When we see that we understand what is happening now as being eternity appearance and getting to know that alive person moving now is a person who lives forever. This is important understanding regarding strategies of future evaluation as well a man, civilization and mankind on the whole.

Having seen strong unlimited level directed toward endless evolution you study first of all constructions of future development systems. Having seen how some cities or some settlements, villages some point of peoples' dwelling in general have been built for instance from the point of view of human being perceiving by the people who have made them you may find out that en element of eternity was laid here. The point is whether to parcel out the earth into nature system and artificial reproduced, artificial reproduced premises include element of eternity related to a human being cognition. The important constituent understandable with regard to the fact that all external worlds' systems the same way are transforming information into a certain reality which a human being may reproduce foe his future evolution. When you are considering this connection logically it becomes obvious that a man takes into consideration presence of air, seas, oceans, the earth according to what he makes construction besides the rest having in mind presence of outer space for instance.

When logical phase of that understanding crosses with task of eternal evolution at the level of having already made objects one may trace connection from all elements of external reality and how they effect on a particular man because of which a particular element of eternity was reproduced anywhere. According to the system of general connections the element has the same eternal evolution as a particular man. Because made by work and made by hands of a man, just as well first move and first thoughts in general, creation of something original including conception it is a reason which gives birth to eternity and which makes eternity either as made by a man himself or by neighborhood people. It is made by means of informational reality and inner apprehension of a man.

Exerted integrity constitutes physical body of a man, structures of external world, bodies of animals, plants and et cetera. Perceivable and apparent at the level of world understanding the inner harmony is an element not covered in words as well as world understanding. Understanding always is very close to eternity of following events. On the basis of this it becomes apparent that everything is eternal. Eternity is obviously clear and it often is not qualified. It often is not perceived at the mental level as something separate from that word itself. That is word as it is as well as any action is not separated from eternity. Therefore when one asks a question whether it possible or not to speak of something using the same at the same time. The example may be used when speaking about sound you use a sound. It is necessary that somebody would perceive a sound. So it is one contemplative example when another man just is an element proved that system is locked and everything has just been realized. A sound is coincident with a message about it. This message is positively informational. This message quotes that en eternity exists. It is displayed in human being actions and in his inner confidentiality of eternal life. Therefore it is possible to select such a constituent that characterized as collective constituent socialized and shaped with the fact that people understand that end of the world and end of physical and spiritual evolution are not going to happen. The end of planet evolution and every separate person for prospect is not going to happen. As if there are two people at least who are just living at the planet so the planet will develop forever. This unity of the two who hands out a signal and who receives it just makes real existence of the Creator.

God the Creator is united in the fact that he can make a signal and can receive it solely. For the earth it means action from t\'VO people. When God is a signal source and an individual who receives any informational signal for him a signal is not divided. Therefore consolidation of the world that is essence of any object is as a matter of fact eternal and unlimited strong. From logical point of view there are possible problems for the earth for instance from asteroids and so on including local problems of civilization development. Actually initial eternity at the spiritual level is like it might not he changed in practice sufficiently not to say about dramatic change the sort of planet drop from its orbit and similar. Phenomenon and events which would impede that will occur always. It is because precisely eternity that defined by two peoples and created the world God the Creator himself, is interpreted adequately like unchangeable uniform eternity. Premised on that one may exactly determine whether one person either may make an impulse similarly how God makes it or also may receive it back.

Therefore thought about eternal evolution and that end of life is not going to happen are the thoughts which returns to a man himself and creates his actual mentality related to eternal life and eternal evolution.

Then people with the aid of faculty of vision will see eternity light which fills these eyes. Their eye lashes will be the sort of so as everywhere at all ends of the world of actuality the penetrated through these lashes light and coming into these eyes images of the world 'Will be seen. Everyone with his own eyes perceives the eternity which fills him. He may so deep enter into actuality that he will just not he identified with actuality but he himself will be eternal reality. This is the message coming into his heart. His heart will become eternal as well as his blood. When he look at himself in the actuality he will understand he will become eternal because he will apprehend like this.

Eternity of mentality is most important. Because sole thought coming to human beings mind never will vanish somewhere. Mentality is always with a person at that level as a prop of personality basis. A person never forget about what he was thinking one day. It is the structure of eternity on which all external world among the rest is building. Confidence towards all the world existence as a result of his mentality is very important from viewpoint of transformation of the world into eternity. Because eternity which fills external world is with what he identifies himself. At that we comprehend that his thoughts penetration at the bottom of essence of his consciousness is what he perceives with regard of himself and with regard of all people. That movement is this beam of inner sun illuminated his movements within actuality. He becomes a real himself as if he gets to know everything. That is an Omni knowledgeable human being. Premised on that actuality of Omni knowledgeable person we see that way of achievability of everything. We see this way is constituted from actual and invented by cognition light. Light of actuality combines with cognition light bringing into world things and time, biological course of events and entity of living matter, and starts to react as consciousness structure. The combination of light of inner and outer actuality of a man ensures that the end of the world is not going to happen, consciousness relates to the first type of lights and the second type leads to living matter creation. It takes for granted also because light of your own thoughts is not affected by external influence since any event saturates thought solely that is makes in more luminous and able to create more living matter in addition to existing. It follows from here the proving that processes of eternity always increase at place of people location. It means that the end of the world is impossible objective processes in the world happening according to laws of development. Whether you see organism life as living of consciousness from you will see that the form is in evolution infinitely at fixed geometrical shape. Soul makes luminescence of a body unlimited. Having a look at that a man starts to see direct connection of his organism with eternal external world.

An actuality appears that embodied the fact of human being's attainment of eternity which is initially in everything surrounding him when he touches any object of external world for instance if he strokes a flower with his hand. He apprehends that infinitely and eternity of external world means as well him among other things. Since a man just understands that and the point of consciousness works out in concrete often mundane activity the man becomes eternal himself at any actuality including mundane, thoughtful, prospective, strategic and so on. It follows from here that actual movement a man to eternity is his personal way which simultaneously fathomed as way of duty of a man. Duty before his entity and before God for moving in eternal evolution. This duty allows making even consciousness structures at the level of current logic. It allows fathoming of occurring in Extraterrestrial. Doomsday is not going to happen civilization on the whole, proper reacting on happened events and thus keeping a balance among external information systems conjugated to consciousness and intercrossing at the level of mentality one way or another. Purpose of the duty conjugated to consciousness is to get integrity of all aspects of personal and precisely eternal evolution beginning with entity of eternal development of each person and consequently identifying additional levels such as duty of eternal development, intention for eternal development, consciousness aside of eternal development and others. Premised on that one may understand that whatever way a person uses to realize his ambitions he has eternal development everywhere. It is his move to eternity which is stable even because everything that arises from a man or from God is eternal therefore they are the same eternal.

On the basis of that level it is possible to identify eternity of God life just according to this factor therefore it possible to identify similarly eternity of a man's life according to the same factor. From intercrossing of these systems one may to move over to a literal principle of eternity realization excluded any end of the world. For instance moving over to letter "X" we study in the case two lines of God and of a man intercrossed in the center of the two light cut-offs. Looking in vision of light of the letter we watch this intercrossing in each letter. For instance in letter "0" one may see upper part of half circle as light element of God and bottom part as a man and surrounding world. When we study according to similar principle all letters we watch in literal terms system that there is no interaction of two elements in word and in information ending's sound systems. It is often enough to know about that for it was fixed that way.

The level of knowledge which fixes eternal actuality according to a will of man is effective method of control. Tool used by a man for life is eternal originally. Therefore life itself does not suppose any end. Sound from expressed in symbols for example is intercrossed with a task and with eternal development actuality that supposed to be facilitated at eternal level of each men's life what he recognizes and moves to proper direction. Sound is what id reproduced for example by the man when we see speaking or singing man in a word man who consciously reproduce certain logical well connected sound system. Clearly word coincides with sound either perceivably for word consisted with letters or within a part where sound intercrosses with a word in unlimited continuity. That is light corresponding to a sound having unlimited continuity for it is light of a man himself connects to word. Word attains unlimited forms and infinite consequences and spreading which are very local initially. Given simple word expressed in from of letters "s 1 o v o" will be here absolutely local being perceived and at sheet of paper. It will be infinite in meaning that was just explained. Intercrossing word and sound coming from a man infinitely within his activity makes new eternal actuality.

From viewpoint of intellectual constituent and original level of creation words are mental activity of a man who invented letters and expressed a word by means of them. Therefore premised on basis of previous generations where the same mental activity took place eternal system displayed at the level of information connections. Perceiving and man's understanding of one external subject of information are as well activity of a man. Therefore eternity by itself within the dynamic meaning arise from a man. Any perceiving it is just eternity of a man. Because law of eternal development that may not be avoided is laid in entity of perceiving of all phenomenon. That law is untouchable essentially from viewpoint of any influence on the law. It is unchangeable in cooperative and public understanding and from viewpoint of its eternal realization. In connection with and premised on that mental constructions which quote of all eternity as a matter of fact is criteria and basis of the world it might unambiguously be said that no end of the world is going to happen as well in physical or in spiritual terms.

Taking into consideration a level regarding to the end having in mind a life of any concrete man it needs to understand that exactly this might not he proposed consciously because general system reacts so as if we exclude concept of death of one person and learn to find out techniques of preventing death of each concrete person, external actuality will fully automatic exclude those conception system of events and symbol forms which are referred to end of the world notion.

Therefore one of system factors of excluding of these global threat for mankind on the whole notions is precise following of death excluding for every concrete person as well as entire mankind. Technique of resurrection here is realized in terms of eternal development. At that living people have to build a system in such manner that there would be no dying of living people. Once this element would be wide spread it will be pretty categorically clear that some whatever information systems emerged on going to the end of the world as meant in cataclysm even in logical phase in general will not appear for controlling of external and internal world would be categorical set up in love.

All this will be united in terms of concrete technical purposes of evolution of society, a concrete person, civilization, worlds and so on. Therefore here it is meaningful a participation in certain educational processes which would inform on that whether people would be oriented for unlimited creation premised on a knowledge of a eternal evolution level. Because in general conscience either of a man or of people on the whole prepares external actuality above all.

When conscience gathers proactive attitude and spirit and soul of a man become pretty collaborative with common eternal evolution level task understanding, it appears very simple educated element on going to whether other people simply realize, admit and disseminate this area of life in the case eternal one. When you would notify others on that then just others may construct informational, actual, practical basis for their eternal development as well as a certain ground for eternal evolution. So your further activity will be more enforced and systematic. Notion of systematic or consistency in eternal development means an eternal life achievement either of a separate person or of mankind on the whole.

Therefore premised on the task, given practice, given life standard one may declare that you come close to eternity because other people are includes in the action too. At that your action is imperturbable just because you move further toward according an eternal evolution level. Reasoning from given practice one may say that more actual results of eternal development and eternal life and each concrete person who learnt or being learning you have seen from eternal evolution viewpoint, more stable is your own eternity having in mind that it is naturally imperturbable. It is desirable to develop massive mundane understanding of the question when eternity appears to be available for wide mass and acceptable for public as well as mundane phenomenon of life. Based upon given level one may consider that knowledge being handed out to other people is precisely systematic eternity included knowledge of eternal development and of eternal life.

Main structure of systematic is just set up at more wide and focused, more comprehensive and more exact level of handing out of knowledge. When you thinking over a handing out of the knowledge you are to think with concentration and you are to realize periodically a system of handing out the knowledge on eternal life, level and development. You will see that handing out of knowledge and effects because of handing out are positive. For example tissues recovery or any systems of events' control information about which was given by other people are a matter that supplies you with certain level of exactness of your position at the level of conscience. Because skill for recovery or improvement of tissues of your own or of other people and skill of influence on certain events in terms of eternal evolution for the according to eternal evolution appropriate situations would take place there, supplies you with current guarantees of eternal evolution at the level of your following actions. %en you acting are consistent and very many people are engaged at that and its obvious that macro-catastrophe is impossible because of internal problems of civilization or of external problems of the world very special evolutionary field of knowledge appears and will set up in future. It possesses a sufficiently very fast rate of information change. Similarly whether to compare quantity of messages in internet it will be very many words however after aggregation of them at the level of electronic signals quantity will be less in terms of the volume for instance such as written by means words according to that signals. Never the less information media and signals themselves are incorporated. Whether to use such structure of world within realization precisely knowledge of eternal evolution, you will see that besides physical fields like internet and the sort of information systems there are a spiritual level of concentrated similar control. Combination at spiritual level of soul of this knowledge that was collected by you and other people for realization of eternal evolution, with those things which we bring into structure of eternal evolution with some data of our own allows influencing on actuality. To dramatically influence at physical reality in a way that become available practically at the level of evens; logic, at the level of certain laws of society evolution that it would be achievable the constructions which will not allow people to die. Self -development of a man allowed to become convinced in possibility to fully influence at the world in creative purposes insisted in end of the world inability. That is combination of events and properties of physical world will be developed in such a way that it will appear not only at the level of purpose or of completion. For instance an airu of a person being in control toward eternal development in the purpose of normalizing of a situation and control itself it will be a standard of eternal life where information processes of change are huge. It is necessary to realize and to create a world in which damage to a health is impossible and it is impossible to terminate people life. Such a world differs with certain properties in terms of spreading of the phenomenon for all as a law which practices in any case. To get round of it or to change is forbidden, because collective conscience has set it up and it is effective as well as physical laws of physical reality. This socialized law of combining with law of inner evolution and laws of external world that is exact practical position which allows not simply be sure that no end of the world is going to happen and as well be sure about this forever. There is an essence of a man which completely is in control over reality. Self-consciousness of a man acting actually at the level of precise reasoning and convincing facts and also at the level of regular system development of society and a man will allow practically to exclude information of such type as the end of the world and as any termination of a human being life under pretense of something. This exclusion just allows to realize structure of eternal development collaborated by a man, external world, society and also God. All system of given development will be directed to eternal life only and it exists right now. Because a man who will be full-fledged in such manner and society in the whole do not differ from people who will own that. Therefore having in mind original spirit and according to God similarity you any act as well as God acts, how another person acts who can develop eternally, who can live forever as it depicted in external reality, where you can independently get to know all laws of eternal development having in mind the purpose referred to external world. When you contact to external world for the purpose this knowledge is coming to you the same way. Therefore everyday intention towards that direction gives you constant knowledge. We improve ourselves materially faster than it may be supposed logically. Then at certain level you may see in some moment an element of eternity is realized at you to that extension that you are living under certain internal structures of consciousness of any events and some possible past events or some your actions which brought with certain results. It is just personal control, personal position which gives certain pleasure and positive perceive of all people are living eternally as well as you.

As a result on the reasoning from given construction of your condition you will be sure that no end of the world is going to happen. You may instantly prevent any demolishing events in any system thanks to this condition, this peculiar attitude where actual strength presents, that is relation built upon precise cognition of future events. With this you set up not only eternity of yourself as well eternity of the world and of all surrounding. When you achieve such harmony and very often it is easy to achieve. You have literally to bestir yourself for that and you will have it. It is such spiritual condition and everything will work. Thus it is absolutely obviously that the world is indestructible about which it might be reported to people close to you and one may have knowledge which may allow you to transmit this at the level of common not different from mundane action state. One may very quickly transmit precisely given confidence and given system of exact knowledge. Then as a result of your action nothing demolishing is going to happen in any case.

Thus with collaboration of understanding, action systems and system simply because of what you are realized as a personality as a man who loves forever with incorporation of all possible phenomenon towards this you obviously will achieve of eternal life without any end of the world or in general speaking any just negative elements which may bring a damage to health or to life.

Translation from Russian into English was made by Permanent Creation.

Other literature by the same author you may like:

EPILOGUE.

Yelena Loginova

I was born in the Altai, Siberia. I have two specialties, chemical production engineer and a nurse in the United States. I earned my nursing degree at Boston College. For five years I worked as the assistant nurse, but since 2005, while working as a nurse, I had a stunning variety of experiences that opened up my eyes to the true state of affairs related to human health, the state of medicine at the moment and cleared all the illusions that I had before. I have also worked in Boston prison for about two years and am very thankful for an interesting experience. I also know that being distributor of Knowledge of Grigori Grabovoi, requires the student at this stage of human development a great inner courage.

I want to transfer my very concrete knowledge that in 1963 Messiah finally came to Earth under the name Grigori Grabovoi. He was born in a village in Asia, as it was accurately predicted by the great seer of the twentieth century, Jane Dixon, in the United States in Washington in 1962, about a year before the birth of Grigori Grabovoi.

I sought Him out consciously since 1992, on the basis of Jane Dixon's prophecy, but only found Master in 2004 already living at that time for six years in the United States. I am a citizen of this wonderful country, and I love America.

I consider myself worthy of the Nobel Peace Prize only because I accurately deciphered several quatrains of Nostradmus from French and English languages (translated myself) and the prophecy of Jane Dixon.

Let me elaborate on this vision and explanation. But one vision of what happened in the early morning in Washington, February 5, 1962, Jane Dixon considered to be the most important in her life.

I would like to state here what exactly that page of the book has revealed to me: "The next morning she overslept, but the sun was not yet up as she walked towards the bay window of her bedroom, which faces east.

As she gazed outside she saw, not the bare-limbed trees and City Street below, but a bright blue sky above a barren desert. Just above the horizon was the brightest sun that she had ever seen, glowing like a golden ball. Splashing from the orb in every direction were brilliant rays, which seemed to be drawing the earth toward it like a magnet. Stepping out of the brightness of the sun's rays, hand in hand, were a Pharaoh and Queen Nefertiti. Cradled in the Queen's other arm was a baby. His ragged, soiled clothing in startling contract to the gorgeously arrayed couple. "The eyes of this child were all-knowing" Jane says softly. "They were full of wisdom and knowledge."

A little to one side of the Queen Nefertiti, Jane could glimpse a pyramid. While she watched entranced, the couple advanced toward her and thrust forth the baby as if offering it to the entire world. Within the ball of the sun, Jane saw Joseph guiding the tableau like a puppeteer pulling strings. Now rays of the light burst forth from the baby, blending with those of the sun and obliterating the Pharaoh from her sight. Off to the left she observed that Queen Nefertiti was walking away, "thousands of miles into the past". The Queen paused beside a large brown water jug and as she stooped and cupped her hands to drink, she was stabbed in the back by a dagger. Jane says that she "distinctly heard her death scream as she vanished".

Jane shifter her gaze back to the baby. He has by now grown to manhood and a small cross, which formed about him, began to expand until it "dripped over the earth in all directions. Simultaneously, peoples of every race, religion, and color, each kneeing and lifting his arms in worshipful adoration, surrounded him. They were all as one". Unlike previous visions, which had gradually faded away from Jean, this one moved ever nearer until she seemed to be in the very midst of the action, joining in the adoring worship. "I felt like a tiny seed ready to sprout and grow". She says, "But I was only one of millions of similar seeds. I knew within my heart, 'Here is the beginning of wisdom'". The room was becoming dark again, and though she was still caught up in the spell of the vision, Jane glanced automatically at her bedside clock. The time was 7:17 A.M.

"What does it mean? What is the significance of this strange vision on a dull February morning in Washington, a third of the way around the world from Egypt?" Jane feels that she has been shown that answer. A bit haltingly, she explains it this way "A child, born somewhere in the Middle East shortly after 7 am on February 5, 1962, will revolutionize the world.

Before the close of the century he will bring together all mankind in one all-embracing faith. This will be the foundation of a new Christianity, with every sect and screed united through this man who will walk among the people to spread the wisdom of the Almighty Power".

This person, though born of humble peasant origin, is a descendant of Queen Nefertiti and her Pharaoh husband; of this I am sure. There was nothing kingly about his coming-no kings or shepherds to do homage to this newborn baby-but he is the answer to the prayers of a troubled world. Mankind will begin to feel the great force of this man in the early 1980's, and during the subsequent ten years the world, as we know it, will be reshaped and revamped into one without wars or suffering. His power will grow greatly until 1999, at which time the peoples of this earth will probably discover the full meaning of the vision.

Attempting to describe her own sensation, Jane says "I felt suspended and enfolded, as if I were surrounded bye whipped cream. For the first time I understood the full meaning of the biblical phrase, "My cup runneth over". I love all mankind. I felt that I would never again need food or sleep, because I had experienced perfect peace".

Jane died in 1997 without having understood who that was, who will save, and change this world. While still alive, she repeatedly was mocked by the press precisely because of this prophecy, because she focused on it her entire life. Even Nancy Reagan, her former friend and wife of former U.S. president, has not regretted it, expressing subsequently apparent disbelief.

I want to quote the words of Jane Dixon, which reveal much to inquisitive researcher: *"Hope of the World, its revival will come from Russia and they have no connection with what is communism. Precisely to Russia the most authentic and a great source of freedom will come... It will be a completely different way of living based on the principle, which will be the basis of a new philosophy of life".*

Prophecy clearly shows that the True Teaching will start its march around the world from Russia, and will be given in the original version, in Russian.

"The Son of Man will come at an hour when you are not expecting him" (Matt. 24.44).

It is of my deep conviction that such a man, the Savior of the world, even as a child, already carries the ability of the Supreme Spirit, phenomenal abilities, as well as a lot of amazing characters. Here is a fragment from the book by Elena Muhovikova, a physician with experience, who met with the mother of Grigori Grabovoi, Lyudmila Ilinichna, and have recorded the conversation.

"Gregory grew up as a quiet boy, never raised his voice on anyone, did not cry, did not fight, and was perfectly healthy. According to his mother Ludmila Ilinichny, he never had a medical card, he was never ill, he did not have any vaccinations, and he never spoke to doctors.

When Gregory was five years old, my parents decided to show him to one clairvoyant old man, Azerbaijanis, who lived in the village "Ilichevka " - 20 km from them. Lyudmila says that: "When we entered his yard, we were met by the son of a clairvoyant who began to say that his grandfather is not receiving anyone, as he is 80 years old, and he is weakened. We have turned around and were ready to leave, but we were stopped by screams of an old man, who demanded to bring Russian boy to him immediately", while parents were asked to stay in the yard. A long agonizing hour has passed of waiting before Grigori appeared before parents. "Grigori, what were you two doing there for so long?" - Asked Lyudmila. "Nothing" - Grigori replied. "He fell to his knees in front of me and kissed first my hand, then my forehead, and then prayed for an hour. Then the old man got up again, kissed my hands, forehead and told me now to invite you to him". When parents faced an old man, he solemnly asked them if they know that their son is born of God? Lyudmila confirmed it with a nod. Old man continued, "Your son, God conceived a child and not a single hair to fall off his head. Keep him as the apple of an eye, and do not reveal him to anyone else. The boy has excellent health. If anyone tries to offend him during his life, then a curse will befall on them for seven generations". Remembering mandate of a clairvoyant, Gregory's grandmother became his main guardian, who accompanied him everywhere.

At school, Grigori was chairman of the head of the Komsomol group. He tried by his own example to educate classmates. Was always calm, never raised his voice on anyone. Always has been perfectly healthy. With absolutely normal mentality, he always sat quietly at his desk, looked no different from the others, and could easily save the world. In fifth grade, Grigori studied the structure of the remote action. Lyudmila told how at Gregory's school, near the main front door, there was a broken window frame with chipped paint, which created an unsightly appearance for the building. Grigori often told his classmate Erken Tazhibayev that a new frame should be installed. Grigori asked: "At lease repaint the frame, if you are not willing to get it repaired". But none of the adults listened to his repeated requests. Then Grigori, for educational purposes, in full view of the children, by a single glance at the frame, remotely knocked the frame out, and it all got warped and with nails sticking out, flew out and hung itself on the tree. For the frame to not fall from the tree and not traumatized children, Grigori had hit it with yet another glance and broke off the branch on which the frame hung. After that, the school administration had to quickly put a new frame in. However, Grigori's grandmother still had to pay for the new frame herself".

Sometime in the future all those who correctly perceived knowledge coming from the World Teacher, those who survived and have not betrayed, did not compromise with his conscience, did not break down in a hail of ridicule even relatives and friends, will receive a great homage. One great philosopher once said, "Do your thing and let people say whatever they like... "But this credo is not for the faint. Not all can resist the collective consciousness, but only very strong people, who do not allow to be manipulated. These altruists are the vanguards of humanity, world servers, who understand clearly Tasks of Evolution. Sounds grand? But such are realities of today's moment: either a global disaster or transition to a new, higher level of development. There is no middle ground.

Media wrote whatever they liked, but not about, for example, the most intelligent people, including many doctors, PhDs, lawyers, teachers, who for a very long time participated in the action which has been declared national, as it has been joined by supporters of Uzbekistan, Ukraine, Belarus, Kazakhstan, Latvia, Estonia, Germany, and the USA. People went on to a three-day hunger strike, people spent replacing each other and sending its notarized listings on holding a hunger strike with demands made to the law enforcement agencies, authorities of the Russian Federation, to the media, embassies and foreign governments. In this unprecedented act participated only about a hundred people. This extreme form of protest, followers of Grabovoi based on the fact that no previously taken measures by them for nearly two years, have not led to any results.

Thousands of letters that were sent to the courts, to the Prosecutor General of the Russian Federation, to the Russian president, and to human rights organizations either went unanswered or stayed "floating in circles", i.e. either got returned to those people who violated the law, or to those to whom complaints were directed to. More than 500 pickets and rallies, that were held in Moscow and abroad, as well as numerous press conferences, were hushed by press.

But thanks to God, the doctrine of eternal and harmonious development is spreading rapidly around the world, books are translated into many languages and are producing phenomenal results that practice Technologies brought to the world by Grigori Grabovoi is the best proof that we are on the right track, on the right way to Eternity.

Yelena Loginova
President of the Research Center of ENLIGHTENMENT. USA

CONTENTS

INTRODUCTION 3

CHAPTER 1
ABOUT THE TEACHING OF
GRIGORI GRABOVOI 5

CHAPTER 2
PROGRESS OF CONSCIOUSNESS. 7

CHAPTER 3
SOME QUOTES FROM BOOKS AND
LECTURES OF GRIGORI GRABOVOI: 9

CHAPTER 4
METHODS OF MANAGEMENT HEALTH, A REALITY 15

CHAPTER 5
RECOVERY AND TREATMENT BY POWER OF WORD 19

CHAPTER 6
SIMPLE GEOMETRIC SHAPES OF INFORMATION 22

CHAPTER 7
A SPIRITUAL ASPECT OF HUMAN 24

CHAPTER 8
WHAT IS CLOSE AND DISTANT
FRACTION OF CONSCIOUSNESS. 26

CHAPTER 9
A FEW EXAMPLES OF TECHNOLOGIES OF
CONSCIOUSNESS. 27

CHAPTER 10
HEALING NUMBER CODES: 36

CHAPTER 11
CONCENTRATION WITH NUMBERS OF SERIES
FOR EACH DAY FOR EVERYDAY
CONSCIOUSNESS DEVELOPMENT 37

CHAPTER 12
A FEW EXAMPLES OF TECHNOLOGIES OF
CONSCIOUSNESS USING NUMERICAL CODS. 59

CHAPTER 13
ADDITIONAL NUMERICAL SEQUENCES 62

CHAPTER 14
RESURRECTION OF GRAIG 66

CHAPTER 15
TEACHING ABOUT ETERNAL LIFE 67

CHAPTER 16
DOCTRINE OF THE EARTH 82

CHAPTER 17
STRUCTURIZATION OF CONSCIOUSNESS.
METHOD OF CONSTRUCTION A DYNNAMIC
ELEMENT OF REALITY IN PERCEPTION. 98

CHAPTER 18
EXTRATERRESTRIAL. 130
DOOMSDAY IS NOT GOING TO HAPPEN 132

EPILOGUE 145

PATH TO ETERNITY

Practical Guidance and Case Study
Applications of Eternal Creative Harmonious Development

Cover design : Jullian

Printed in United State of America

ENLIGHTENMENT PUBLICATIONS

Printed in Dunstable, United Kingdom

91221339R00087